THE COMPLETE IDIOT'S GUIDE® TO

Open Nesting

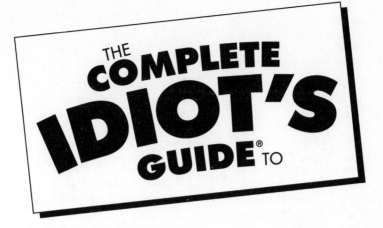

THE COMPLETE IDIOT'S GUIDE® TO

Open Nesting

by Lauren A. Gray, MS, LMFT, and
Wendy Bedwell-Wilson

ALPHA

A member of Penguin Group (USA) Inc.

This book is dedicated to my mother, Marilyn, and father, Bill, who have been wonderfully supportive throughout my life—even when I needed to open nest! It is also dedicated to my baby daughter, Ánja, who has shown me how strong a parent's love can be.

—Lauren

ALPHA BOOKS

Published by the Penguin Group

Penguin Group (USA) Inc., 375 Hudson Street, New York, New York 10014, USA

Penguin Group (Canada), 90 Eglinton Avenue East, Suite 700, Toronto, Ontario M4P 2Y3, Canada (a division of Pearson Penguin Canada Inc.)

Penguin Books Ltd., 80 Strand, London WC2R 0RL, England

Penguin Ireland, 25 St. Stephen's Green, Dublin 2, Ireland (a division of Penguin Books Ltd.)

Penguin Group (Australia), 250 Camberwell Road, Camberwell, Victoria 3124, Australia (a division of Pearson Australia Group Pty. Ltd.)

Penguin Books India Pvt. Ltd., 11 Community Centre, Panchsheel Park, New Delhi—110 017, India

Penguin Group (NZ), 67 Apollo Drive, Rosedale, North Shore, Auckland 1311, New Zealand (a division of Pearson New Zealand Ltd.)

Penguin Books (South Africa) (Pty.) Ltd., 24 Sturdee Avenue, Rosebank, Johannesburg 2196, South Africa

Penguin Books Ltd., Registered Offices: 80 Strand, London WC2R 0RL, England

Copyright © 2009 by Amaranth IlluminAre

International Standard Book Number: 978-1-59257-938-9
Library of Congress Catalog Card Number: 2009924928

11   10   09   8   7   6   5   4   3   2   1

Interpretation of the printing code: The rightmost number of the first series of numbers is the year of the book's printing; the rightmost number of the second series of numbers is the number of the book's printing. For example, a printing code of 09-1 shows that the first printing occurred in 2009.

Printed in the United States of America

Note: This publication contains the opinions and ideas of its authors. It is intended to provide helpful and informative material on the subject matter covered. It is sold with the understanding that the authors, book producer, and publisher are not engaged in rendering professional services in the book. If the reader requires personal assistance or advice, a competent professional should be consulted.

The authors, book producer, and publisher specifically disclaim any responsibility for any liability, loss, or risk, personal or otherwise, which is incurred as a consequence, directly or indirectly, of the use and application of any of the contents of this book.

Publisher: *Marie Butler-Knight*
Senior Editorial Director: *Mike Sanders*
Managing Editor: *Billy Fields*
Executive Editor: *Randy Ladenheim-Gil*
Book Producer: *Lee Ann Chearney/ Amaranth IlluminAre*
Development Editor: *Lynn Northrup*
Senior Production Editor: *Megan Douglass*

Copy Editor: *Jan Zoya*
Cover Designer: *Becky Harmon*
Book Designer: *Trina Wurst*
Indexer: *Johnna VanHoose Dinse*
Layout: *Ayanna Lacey*
Proofreader: *Mary Hunt*

Contents at a Glance

Part 1:   Going from Empty Nest to Open Nest   1

   1 There's No Place Like Home   3
*What is open nesting, and why are parents and
their adult children doing it? In this chapter,
we explore reasons why a family may choose to
share space again.*

   2 Tough Times Call For Bold Moves   13
*Both parents and adult children will see benefits
and drawbacks of open nesting. We'll outline
some common pros and cons, which will help
you decide whether living together is a viable
option for you … and for your family.*

   3 A New Way to Be Independent   25
*As we progress through life, we accomplish spe-
cific developmental tasks or goals that help us to
function well as individuals and as members of
a family. In this chapter, we discuss these tasks
so that you can be sure every member of your
family is getting what they need for the stage
they're in.*

   4 We Are Family   39
*What are your family dynamics like? What
is "healthy" or "functional"? Using the
Circumplex Model, we'll introduce different
ways family members tend to interact—and
which styles work best in an open nest.*

Part 2:   Living Together   53

   5 Stay Awhile … Stay Forever!   55
*How long is too long? In this chapter, we'll
examine how to determine the length of time
your child should remain in the nest. We'll also
present common challenges that may threaten
to derail your timeframe, and discuss how to
overcome them.*

6 Making Room   69

*Perhaps your son's room hasn't changed since he
left the first time. Maybe you've turned your
daughter's room into a den. In this chapter,
we'll walk through the house and look at ways
to make physical room for your adult child—
since your old arrangement may no longer fly.*

7 Three's Company ... Is Four a Crowd?   79

*What happens if your child moves home with
her or his married partner (and even a child)?
The dynamics in the home will certainly change.
We'll offer advice to make the two-couples-
under-one-roof scenario work for everyone.*

8 Dating and Sleepovers   89

*Intimacy and sexual relations are normal and
natural for a young adult, but what does that
mean for parents whose adult child is living at
home? Is sex a taboo topic in your home? In this
chapter, we'll look at common issues that may
surface if your child is in an intimate relationship.*

9 Chores Are Boring, Rent Is Rude   103

*Because your child is now an adult, you can
expect your son to help around the house or
your daughter to pay rent or pitch in for
utilities. We'll look at ways to divvy up labor
in your open nest, and help you decide whether
to charge your child rent.*

Part 3:   **Rules? Who Needs Rules?**   113

10 Whose House Is This, Anyway?!   115

*Your open nest is a group of adults—so who's
supposed to be "in charge"? In this chapter,
we'll look at how important decisions will
be made. We'll also examine issues around
emotional closeness, and how involved family
members "should" be with one another's lives.*

11 Day-to-Day Life Around Here   125
*Sometimes, conflicts are unavoidable. Family
members may disagree about chores or personal
space. We'll offer tips and techniques for learning how to communicate with each other and
handling these day-to-day squabbles.*

12 Sisters and Brothers   137
*Every family member deserves a chance to take
the spotlight, so we want to focus the lens on the
sisters and brothers of adult children who move
back to the nest. We'll talk about how to give
those children the attention they need to grow
and thrive and how to keep them from living
in the returning child's shadow.*

13 What's for Dinner?   147
*Does everyone in your house eat different foods
at different times? Does anyone cook? We'll
talk about what it means to eat together, and
how to create and nourish healthful eating
habits.*

Part 4:   **What, Me Worry?**   **153**

14 Holding On to Your Money   155
*You worked hard for your money. When your
son or daughter is strapped for cash, you may
feel you need to pay your child's way again.
Not so. In this chapter, we'll look at ways to
secure your money—and teach your child how
to manage a bank account.*

15 You Look Too Young to Have Grandkids!   165
*Can you believe your grandchild is living
with you? In this chapter, we explore the
grandparent/grandchild relationship and
discuss what is appropriate and optimal when
your adult child moves back with her children
in tow.*

16 Guess Who's Coming to Dinner ...
    and Staying?   175
    If a child's significant other differs in race,
    ethnicity, political viewpoint, religion, or sexual
    orientation, that can challenge families to find
    new ways to grow together. In this chapter,
    we'll look for strategies to prove true that what
    unites us is greater than what divides us.

Part 5:   Learning to Move On   185

17 Making a Plan   187
    As we say throughout the book, you need a plan
    for how your child's stay with you will go and
    how he or she plans to re-launch into the world.
    In this chapter, we'll suggest 10 essential ele-
    ments to consider when crafting a plan with
    (and for) your adult child.

18 When They Keep Coming Back   199
    Is this the second (or third or fourth) time your
    adult child has moved home? There may be
    a perfectly logical explanation for the repeat
    roommate—or it could mean that your child
    needs more extensive help. We'll look at com-
    mon reasons why an adult child keeps coming
    back and help you decide what's reasonable.

19 Planning an Escape ... *Yours*   209
    You had plans and dreams for your own life,
    and there's no reason to put those on hold while
    your adult child is home. In this chapter, we'll
    offer strategies to keep you on track with your
    personal goals.

Appendixes

A What Children Will Agree To: A Contract   217

B What Parents Will Agree To: A Contract   221

C Resources   223

   Index   227

Contents

Part 1: **Going from Empty Nest to Open Nest**   **1**

1 **There's No Place Like Home**   **3**

A Fresh Perspective 4

A Parent's Perspective 6

Why the "Vacancy" Light Is On 7

Planning Makes the Difference 8

Mom, Dad: I'm Home! 9

Survey Says 10

Red Flags.. 11

Homeward Bound 12

2 **Tough Times Call for Bold Moves**   **13**

Following Your Gut Instinct................................... 14

Examining Your Personal Perspective..................... 14

Making a Pro-Con List.. 16

Parents' Pros and Cons 17

Benefits of Open Nesting 17

Drawbacks to Open Nesting 19

Pros and Cons for the Adult Child......................... 21

Benefits to Moving Home................................... 21

Drawbacks to Moving Home 22

3 **A New Way to Be Independent**   **25**

What's Good for the Individuals 26

The Young Adults' Life Tasks 27

The Parents' Life Tasks...................................... 28

What's Good for the Family 29

Establishing Independence 30

Forming Healthy Connectedness 31

Are We Differentiated Yet? 32

Treating Them Like Adults................................. 34

Setting and Achieving Your Goals 34

Challenges in Process... 35

Goal-Setting in an Open Nest............................. 36

Goal-Setting Exercise.. 37

4 We Are Family   39

The "Traditional" Family...40
Modern Configurations 41
   Single/Divorced Parent..*41*
   Remarried With or Without Kids at Home *42*
   Multigenerational Household *43*
Understanding How a Family Functions 44
   Cohesion *44*
   Flexibility *46*
   Communication *47*
Family Dynamics and the Open Nest 49
   Feeling Like a Third Wheel.....................................*49*
   Younger Siblings Displaced.....................................*50*
   A Return to Old Roles...*51*

Part 2: Living Together   53

5 Stay Awhile ... Stay Forever!   55

Mismatched Parent-Child Goals..................................56
   The Analysis..*57*
   The Solutions...*57*
Conflicting Goals Between Partners 58
   The Analysis..*59*
   The Solutions...*59*
Your Child's Not Ready to Leave 60
   The Analysis..*61*
   The Solutions...*61*
Parental Road Blocking...64
   The Analysis..*65*
   The Solutions...*65*
Parental Pity 66
   The Analysis..*66*
   The Solutions...*67*
A Dream or a Nightmare? 67

6 Making Room   69

Give Me Space..70
Then Versus Now 71
The Bedroom 72

The Kitchen 73
The Living Space 75
Sharing Technology... 76

7 Three's Company ... Is Four a Crowd?   79

Conflict Resolution 80
  A Failure to Communicate *80*
  And Junior's Partner Makes Four....................... *81*
Blending Well—or Not... 83
"Nesting" in the Nest 85
Get a Room! 86
Parents Who Won't Let Go 87

8 Dating and Sleepovers   89

Are You OK With It? 90
A Taboo Topic?... 91
Accept It! 92
  Sleepover Stress.. *92*
  Bringing Partners Home *94*
  An Unacceptable Partner *95*
  Revolving Door of Partners............................... *96*
  What Younger Siblings See *98*
Dad or Mom Dating 98
  Guilt About Abandoning an Adult Child............... *99*
  Privacy, Please!.. *99*
  Your Child Disapproves of Your Having Sex............ *100*
  Your Child Disapproves of Your Dating................. *101*
Communicate! 101

9 Chores Are Boring, Rent Is Rude   103

Who Does What Around the House?...................... 104
Challenges with Chores 105
  Regressing into Old Roles *105*
  Your Child Is Not Doing the Job *106*
Rent? Really? 108
When Rent Goes Bad 109
  Not Paying the Rent.. *110*
  New Rent Rules.. *111*
Take It as It Comes 111

Part 3:   Rules? Who Needs Rules?   **113**

10   Whose House Is This, Anyway?!   **115**

Making Adjustments 116
Who's in Charge? 117
  A Rigid Family *118*
  A Chaotic Family *119*
  Remember: You're in Charge *120*
Gauging Your Distance 121
  Family Is Too Enmeshed *122*
  Family Is Too Disengaged *123*

11   Day-to-Day Life Around Here   **125**

Typical Tensions and Conflicts 126
Communication Breakdown 127
  Active Listening *128*
  Using "I Statements" *129*
  Be Mindful of "Mind Reading" *130*
  Ask Nicely *130*
  Strike When the Iron's Cold *131*
  Try "Repair Attempts" *131*
  Acceptance and Change *131*
Chasing—or Running from—Conflict 132
  Take Frequent Breaks *134*
  Use Successful Communication Techniques *134*
  Set Up Weekly Meetings *135*
  Letter Writing *135*
Keep It Positive! 135

12   Sisters and Brothers   **137**

The Sibling Relationship 138
  Age Differences *138*
  Birth Order Differences *139*
Sibling Relationships Through Time 140
  As Adolescents *140*
  As Young Adults *141*
Sharing Resources and Shifting Roles 142
Siblings in the Open Nest 143
  Between Launched Siblings *143*
  One Home, One Launched—and Back *144*
  Between Step- or Half-Siblings *146*

13 What's for Dinner?   **147**

You're Buying! 148

Menu Planning and Preparation.......................... 148

How Is It Enjoyed? 151

Resolving Common Issues 151

Part 4:   What, Me Worry?   **153**

14 Holding On to Your Money   **155**

Money Management 101 156

Retirement Ahead—Ready or Not!...................... *156*

Exorbitant Expectations *157*

Money in the Family... 157

Family History *158*

Teach Them Well.. *158*

Finance Issues in the Open Nest.......................... 159

Bailing Them Out.. *159*

Enabling Entitlement.. *160*

Couple Tension *162*

Keeping Them Under Control................................ *163*

15 You Look Too Young to Have Grandkids!   **165**

Grandparenthood Today.. 166

A Grandparents' Role in the Open Nest................. 168

Discipline and Feedback *168*

Babysitting Duties.. *169*

When Grandparents Raise Grandchildren 171

Clarify Who's Who.. *171*

Give the Child Time to Adjust *172*

Life with Your Grandchild..................................... 173

16 Guess Who's Coming to Dinner ... and Staying?   **175**

Forbidden Subjects? 176

Racial Tolerance 177

Let's Talk Politics 178

Your God, My God, Our God.............................. 180

Who Do You Love? 181

Accepting the Differences..................................... 183

Part 5: Learning to Move On   **185**

17 Making a Plan   **187**

Look at Your Family..188
How "Official" Should Your Plan Be?......................189
Planning for the Stay......................................190
   Issues to Consider....................................*190*
   A Productive Process..................................*192*
Planning for the Re-Launch 193
Family Meetings 101.......................................194

18 When They Keep Coming Back   **199**

An Age or a State of Mind? 200
What's Reasonable?..201
Times of Transition..202
   A Bumpy Ride..*202*
   I'm Back—Again..*203*
   Lost Path...*204*
Trouble Planting Roots....................................205
Emotional Support...206
   Your Child Needs You..................................*206*
   You Need Your Child...................................*207*
One More Time 208

19 Planning an Escape ... *Yours*   **209**

Issues That May Surface and How to Solve Them . 210
   Heading in the Wrong Direction........................*210*
   Trouble "Getting into Character"......................*211*
   Consumed by Guilt.....................................*213*
Staying on Track..214
When They Launch for Good 215

Appendixes

A What Children Will Agree To: A Contract   **217**

B What Parents Will Agree To: A Contract   **221**

C Resources   **223**

Index   **227**

Introduction

Open nesting is when parents welcome an adult child back into their home—an idea you may be very familiar with if you picked up this book! For myriad reasons, from financial difficulties to life changes and everything in between, a young adult may simply need to return to the comfort, support, and safety of family, of *home*.

When your adult child moves back in, your family will need to make some big adjustments. You'll need to reexamine the roles in your family. You'll need to make space, both physical and emotional, in your home. Each family member will need to learn how to be independent while living under the same roof. And most important, everyone will need to work together to achieve a common purpose of seeing your son or daughter succeed.

Granted, it will take some work, but don't worry! We're here to help! Besides Lauren's clinical experience in helping others in your position, we've actually experienced open nesting in our own lives. Lauren lived with her parents for nine months prior to returning to school for her Master's degree. Wendy lived with her parents and in-laws for brief stints when relocating to different parts of the country. We both know adult children who have moved back home, as well as parents who have opened their nests. We've seen all the angles of open nesting, and we're here to offer our wisdom, insight, and advice to help you grow closer as a family.

In this book, we'll walk you through open nesting, from deciding whether to open your home to seeing your child fly from the nest (again). We'll examine how your family might function in an open-nesting environment. We'll explore the ins and outs of living together. We'll also suggest ways to make the new roommate situation benefit the entire family, so that all members can continue to grow and achieve their personal goals.

Hold on tight, stay positive, and look ahead to the opportunity to get to know your child as an *adult*. You'll have a front-row seat watching your son or daughter mature. As Winston Churchill once said, "The optimist sees the opportunity in every difficulty."

About This Book

The Complete Idiot's Guide to Open Nesting is divided into five parts that cover everything you need to know about opening your home to your adult child.

Part 1, "Going from Empty Nest to Open Nest," introduces you to the basic principles of family dynamics and acquaints you with concepts we use throughout the book. You'll discover useful methods for deciding whether your family is ready to live together again in the first place.

Part 2, "Living Together," delves into what it actually looks like to live together again. How will you make space? How will your family divvy up the chores? This practical, problem-solving section of the book will address common issues that come up in day-to-day life.

Part 3, "Rules? Who Needs Rules?" centers on interpersonal relationships, and how to solve conflicts and challenges that may arise between family members, including helping younger siblings keep their hold on the spotlight. You'll learn how to examine the way your family interacts, and explore ways to make the open-nesting environment work.

Part 4, "What, Me Worry?" focuses on ways to stay positive and focused on your goals. As you'll learn, it's critical that you, as parent (or grandparent), move ahead and not regress into a "Mom" or "Dad" role. We help you look at the open-nesting environment in a positive light, as a way for the entire family to succeed together.

Part 5, "Learning to Move On," explores what to do after your adult child launches—for good. What are your life goals? What dreams do you and your partner want to fulfill now? What do you want the next chapter of your life to look like? We'll offer ways to embrace this exciting—and child-free—part of your life.

Tips and More

Throughout each chapter in this book, you'll find three types of extra information, conveniently packaged into little boxes for your reading pleasure. These add to your practical knowledge, telling you anecdotes drawn from typical family experiences, offering little tidbits of information, and helping you out with things you can do, or things you shouldn't do, to make the open-nesting experience as pain-free as possible.

Ground Rules _____

These boxes are full of advice and tips you can use to make your life—and your adult child's life—easier while living together.

Under Your Roof _____

These are the things to watch out for, to help you avoid and solve any problems you might have as you learn to live with your child again.

A New Generation

We've got a lot of stories, factoids, and tidbits, and we'll share them with you here. Look for interesting statistics, cool quotes, and reflections on the younger generation.

Acknowledgments

Lauren Gray: I am grateful to Wendy Bedwell-Wilson, my co-author. Not only is she talented, but also tenacious—her hard work has made this book possible! Her ability to wrap her mind around new concepts and run with them is nothing short of impressive. Thanks to my husband, Josh Hutto, who is behind me in everything I do. Thanks to my wise and kind family therapy mentors, Julie Wood and Nancy Burgoyne. Most of all, thanks to my clients, who have bravely shared their stories with me over the years, and have taught me far more than I could ever teach them.

Wendy Bedwell-Wilson: Thank you to my brilliant, encouraging, and patient co-author, Lauren Gray, whose wisdom gives this book the authority and foundation to help open nesters (and their adult children) in need. Thanks to Lee Ann, for entrusting me with this project, which I enjoyed tremendously. Thanks to all the open nesting families who graciously shared their stories with me. Thanks to my impossibly supportive husband, Ryan, for enduring my deadline stress and truly believing in me. And thanks to my favorite dog, Pete, for making me laugh when I needed it the most.

Trademarks

All terms mentioned in this book that are known to be or are suspected of being trademarks or service marks have been appropriately capitalized. Alpha Books and Penguin Group (USA) Inc. cannot attest to the accuracy of this information. Use of a term in this book should not be regarded as affecting the validity of any trademark or service mark.

Part 1

Going from Empty Nest to Open Nest

Maybe you cried when you put your daughter on Amtrak to New York City with a suitcase to start her career. Or you held your breath when your son bought a one-way ticket out west after college. And you held it again when he signed up to serve his country and shipped out halfway around the world. You knew it when you walked her down the aisle, perhaps, or when you held his first child in your arms. *That's it*, you thought, *the nest is empty*. But wait … they're baaaackkk! In this part, we'll take a first look at all the reasons why you might want to open your nest to your adult daughters and sons, and help you get started. However the saying goes, your adult children really *can* (and do) come home again.

There's No Place Like Home

In This Chapter

- ◆ Your children are flocking home
- ◆ Offering an expert's unique perspective
- ◆ Understanding "open nesting"
- ◆ A parent's point of view
- ◆ A young adult's point of view

If you've picked up this book, chances are you've added a new roommate to your home: your adult child. Ready or not, more and more families are opening the nest. In record numbers, you are welcoming back your sons and daughters, and here they come!

Maybe your 22-year-old chose to follow his dream of becoming a photographer or actor. Perhaps your divorced 25-year-old is rebuilding her life but needs your financial and emotional support. Or maybe your 18-year-old simply hasn't found his calling— or flown the nest yet. It's a phenomenon that's not all that uncommon. Chances are pretty good that you know at least one friend or colleague who has opened his nest, too.

The number of young adults flocking home is staggering: according to the United States Census figures from 2007, 55 percent of men and 48 percent of women ages 18 to 24 and 14 percent of men and 9 percent of women ages 25 to 34 live with one or both of their parents. And it's likely more and more kids will be flying back to the nest, too, as the economy softens and families see the value of coming together to weather financial storms.

Welcoming home your son or daughter presents a wonderful opportunity to get to know your child again—this time, as an adult.

A Fresh Perspective

Before we begin, it's important to introduce our expert author, Lauren Gray. She's a licensed family therapist experienced in coaching families through a variety of issues, including open nesting. But unlike many family therapists, she's closer in age to your son or daughter than likely she is to you, and she has first-hand experience returning home as an adult to live with her folks. She can relate to parents on a professional level while genuinely empathizing with adult children's point of view. She bridges the generations, and it's her young yet wise perspective that makes this book unique.

Lauren has worked as a family therapist since 2002. She helps scores of families improve their relationships with one another and become a stronger family unit. When Lauren works with clients, she examines key family-dynamics issues including the degree of separateness and togetherness between family members, who's in charge of the family at what times, the family's cultural background, treated or untreated mental illness, how the family deals with conflict, and where the family is in their family life cycle or the different emotional and intellectual stages a family goes through as they go through life together. After she assesses the situation, she designs a treatment plan to address any problems.

Even more to the point, Lauren also works with a number of families that are going through the same thing you are: opening their homes to their adult children. While counseling these families, a number of major questions commonly arise—questions you can probably relate to:

◆ Who calls the shots when everyone around is supposed to be an adult?

◆ How much involvement will everyone want to have in each other's lives? What levels of closeness and distance will work for all family members?

◆ How do you view your adult child moving back home? Is it a welcome event or an intrusion? Is it a reasonable, understandable step, or a result of your child's failure (or perceived failure)?

◆ What do you expect in return for housing your child? What are the conditions or stipulations for him continuing to live at home?

◆ How free will your child feel to do as she pleases?

◆ How will the child become equipped to return to independent living?

Besides her clinical experience, Lauren has firsthand knowledge of what it's like to move back home. At age 24, after a stint working in a London-based corporation, she returned to her parents' house to regroup and apply to graduate school. For nine months, she lived in her childhood home. She and her folks spent quality time together—including many long dinner conversations. When her Dad dropped her off in Evanston to start school the following fall, she knew that she and her parents had grown closer.

Ground Rules

When an adult child moves home, some big adjustments will need to be made by all family members. Honestly, it's unrealistic to expect that every family will become closer through an open nesting experience, like Lauren's family did. This book will hopefully help you and your child have a positive experience of living together again—and help families who are thinking of giving up get through the experience.

As you can see, Lauren understands what it's like to move back home as an adult. She also understands how that situation can affect a family's dynamics. Her clinical skills will give you the techniques to work through the open-nesting process; her age and experience will give you a glimpse into what your son or daughter will be going through. She's in a unique position to provide you guidelines for success.

Moms and dads today have become an integral part of their kids' lives—from childhood through adulthood. The whole family is working toward a shared purpose, whether it be launching the child into a career, helping to raise a grandchild, paying off debts, or saving for a down payment for a home of her own. The family is staying together until that goal is achieved.

> ### A New Generation
>
> After an unhappy stint at a local college, 20-year-old Amy decided that her real dream was to become a cosmetologist. She applied to a top salon's training program—and was accepted. Amy felt ecstatic about moving toward her new career, but she had a tough financial decision to make: if she dropped out of college to participate in the training program, she would lose the financial aid that enabled her to live independently in an apartment. After considering the alternatives, Amy decided to move back in with Mom and Dad while completing her intensive 18-month program. Lucky for Amy, her parents welcomed her with open arms.

A Parent's Perspective

Opening your nest to your adult children could be a joy, but it will also be rife with challenges. Maybe you planned a cross-country jaunt in your R.V. after Junior moved out. Perhaps you've turned his bedroom into that state-of-the-art den you've always wanted. If he's been on his own for a while, it's likely you've settled into your own routine, which will need some adjusting. He's probably settled into his own routine, too, which will require adjusting on *his* part.

Why the "Vacancy" Light Is On

So why are Mom and Dad willing to open the nest for their young adult children? You probably know how to answer that question, but experts point out these possible reasons:

◆ Parents today have more discretionary income. Due to successful lifelong careers (and perhaps the remnants of once-strong investments), many middle- and upper-class families enjoy a healthy income and nest egg. They can afford to welcome their children home and, in some cases, support them financially.

◆ A shift from generations before, parents today have more egalitarian bonds with their children. They enjoy having their children around as long as they're pursuing a path they can endorse. Many times, this involves a low-paying career choice, like music or art, but it can also mean supporting them through graduate school.

◆ Some parents have a psychological need to keep their children home. Parenting gives them a sense of purpose or enables them to avoid facing their own marital issues.

◆ Throughout the 1980s and beyond, sociologists note that parents and their 20-something offspring enjoyed a closer bond than previous generations, suggesting that welcoming their child back reflected their close-knit relationship rather than a failure to launch into adulthood.

◆ If parents have to choose between paying for their son's dorm room or apartment and letting him live at home, many moms and dads would prefer the latter. And forking out funds for a down payment on a home of his own—though a wise financial move—may not be feasible for many parents, especially during softening economic times and tightening credit.

> **Under Your Roof**
>
> In this new living situation, privacy rules. Resist any temptation to snoop on your adult child's Facebook or MySpace page, or insist that she change its content.

Planning Makes the Difference

Before you even start thinking about whether to welcome your son or daughter home, consider how the move will affect you and your spouse. For some couples, the prospect of a child coming home will be a dream and strengthen the family as a whole. For others, it could cause a rift greater than the Grand Canyon. In your decision-making process, you'll need to make time for a serious discussion and clearly determine what you each expect.

You'll also want to discuss what your and your spouse's expectations are of the child. You should both be on the same page when discussing the specifics of your and your child's arrangements well before your son or daughter moves back in!

When doing your pre-planning, keep this list of questions handy:

- Are you both willing to welcome your adult child home?

- Will you be able to resist putting your life on hold when your adult child returns?

- Is your relationship with your child a positive one? Will the experience benefit everyone involved?

- Will you unconsciously be using your child as a buffer in a troubled marriage?

- How long will you allow your son or daughter to live with you?

- In what ways (if any) will you support your child financially?

- Will you require him to pay rent? Chip in for bills and groceries? Cook meals?

- Rather than have your child move back in, are there ways you can help her continue to live independently? For some families, this can be an easier option.

- Is there a possibility you are still hoping to "parent" your child by changing her or teaching her? Your child is grown up, and the window for you changing her is all but closed. Count on your child staying as is for the time being. If she changes for the better, you can consider it a bonus!

◆ Are you willing to allow your child to enjoy similar freedoms to what he grew accustomed to when living independently? If not, there will be guaranteed conflict!

◆ How do you and your spouse plan to handle disputes with your child? It may be your house, but you won't have as much leverage with your child as you used to. Disputes are more likely to take the form of negotiation rather than you laying down the law.

There's no doubt that allowing your child to move back into your home will cause stress and anxiety. You've been looking forward to the freedom, to spending alone time with your spouse, to pursuing endeavors you've been dreaming about for years. It's understandable to be anxious about how your child's return will affect your household.

One critical factor in handling your anticipation is how you interpret your child's presence. Is he unemployed and mooching off your generosity or struggling through a bitter divorce? Is she working or going to school to further her career? Your child's reason for being there will directly affect your response to her presence.

If you approach your new situation with a positive point of view, it can be a truly wonderful experience for both you and your child. It's an opportunity to take part in your son or daughter's re-launch into the world.

Mom, Dad: I'm Home!

There's a big wave of young adults flocking back to the nest, and it's only going to get bigger. Experts point to many reasons why this tsunami of sons and daughters is looking to home for support. Populations today are living longer, which has stretched adolescence well into the 20s. Adult children delay moving out or return to the nest due to myriad financial concerns. They're attending post-secondary education in greater numbers, putting off marriage, and focusing on career before starting families of their own.

Survey Says ...

In case you need some hard evidence proving just how tough kids have it these days, check out these facts and figures pulled from the Bureau of Labor and Statistics and the 2007 U.S. Census:

◆ Young adults are used to a higher standard of living. Those in middle- and upper-middle-class households grew up comfortably, to say the least. Trying to duplicate that standard of living isn't easy, especially given today's economic climate.

◆ Paychecks, when adjusted for inflation, are down across all industries. A young adult living in the San Francisco Bay area, for example, needs to bring in an annual income of nearly $80,000 to meet day-to-day expenses. It simply costs more money to live in today's market.

◆ Jobs are disappearing. Reports showed a steady decline in job creation as of late, a trend that may continue going into the next decade.

◆ Skill requirements are up. No longer is a GED, high school diploma, or Bachelor's degree sufficient to win a competitive job in today's market. For many career seekers, the more advanced your degree, the better your chances will be of finding employment—which means more years in school pursuing a Master's degree, a Doctorate, specialty trade certifications, and more.

◆ Throughout the country, rent is costly and prospects of home ownership are often out of reach. For instance, in Chicago a young person can expect to spend more than $1,000 a month to rent a one-bedroom apartment or pay more than $260,000 for a home. It's easier to have your child move home than to gift her a down payment.

◆ Americans spend a lot of money, and many young people have racked up boatloads of debt—credit card debt, student loans, personal loans, car loans—the list goes on and on. Many times, it's easier for them to move back in with Mom and Dad and pay off those bills rather than spend hard-earned money on rent and utilities.

- The average age of marriage keeps rising, hitting 27.5 years old for men and 25.9 for women, according to the U.S. Census. As a result, people in their 20s are experiencing a longer period of being alone, being single—and trying to live off one income.

- Climbing the corporate ladder takes a lot of work. Many young people would rather live at home and focus on their career without the burden of supporting a growing family.

- With the divorce rate at 36.6 percent among women and 38.8 percent among men aged 20 to 24 years old according to the U.S. Census Bureau, many divorcees have no choice but to move home just to get themselves back on their feet again or for emotional support.

- Still another group of young people just haven't left the nest. A young man may be going to school at a nearby community college or state university. A young woman may have a job that enables her to stay at home. They like the comforts of home.

As you can see, it's clear why young people today may need a little help from their folks. They feel they can go to their parents for support and guidance, especially if their relationship with them is a positive one.

Red Flags

While it's true that there are plenty of motives to move back in with Mom and Dad, some adult children may take advantage of overly generous parents. Before you put out the welcome mat, consider these caveats that should make you cautious about your son's or daughter's motives for coming back home:

- If your adult child is just hanging out and has no plans for the future, consider it a warning sign. A young adult with no goals could wind up to be a permanent fixture in your home.

- Another warning sign is if your adult child is dragging her feet when looking for a job or registering for school. Procrastination—for a little while—is one thing. But when weeks or months go by with no job prospects, or semesters come and go with no credits earned, you can pretty much expect that kind of motivation (or lack thereof) to be present when your child moves home.

♦ If you and your child had a strained relationship in the past and there's no sign that things have changed, you may want to rethink inviting him as a roommate. Similarly, if your child didn't get along with his sibling who still lives at home, you may want to take a look at their dynamics.

♦ Excessive partying or drug use, staying out until the dawn's first light, or inviting unwelcome guests into your home may give you pause when considering whether to let your child move back in.

♦ Is this the third, fourth—or more—time your child has flown back to the nest? If so, he may be taking advantage of your good will and you may want to reconsider letting him back.

Homeward Bound

Families like yours are grappling with the idea of living together after the children reach adulthood. For some, it's a no-brainer: parents and children will fall into their comfortable roles and work together toward a common goal. For others, it's a struggle.

Whether your child is embracing home with the fervor of young Dorothy in the classic *Wizard of Oz* who learns to look no further for happiness than her own backyard, or is dragging her young feet to your front door consider it an opportunity to grow as a family. After all, there's no place like home, right?

The Least You Need to Know

♦ More and more families are embracing the open-nesting concept, thanks to increased financial concerns and a closer familial bond.

♦ Though it may seem like a temporary inconvenience, parents may benefit from welcoming their children home—as long as they're prepared for it.

♦ In an increasingly challenging financial climate, adult children may benefit now more than ever from their parents' generosity.

♦ Not every open-nesting situation will be successful. Parents who feel their adult child may take advantage of their generosity may want to reconsider opening the nest.

Chapter 2

Tough Times Call for Bold Moves

In This Chapter

- ♦ Families are weathering today's storms together
- ♦ Following your instinct first
- ♦ Understanding your personal lens
- ♦ Open-nesting benefits and drawbacks for parents and adult children
- ♦ Starting to form a plan

Today's economic climate is tough, to say the least. As we saw in Chapter 1, adult children (and their parents) face myriad challenges as they head out on their own. Increasing credit card and student loan debt, the high cost of living, rising numbers of foreclosures, and job cutbacks mean that families need to make some bold moves to survive. Many are seeing the benefits of coming together under the same roof to weather the brewing storms.

Through these challenging times, many families are indeed surviving—and even thriving. The ones who are doing so take time to think through the new living arrangements before diving in. They follow their instincts. They consider the benefits and drawbacks, formulate a plan, and move forward with specific goals in mind.

Following Your Gut Instinct

Sometimes you just have a "feeling" about a situation. When you first met your spouse, you may have "known" he was the one. When you interview someone for a job, you may get a good—or bad—feeling about her. In his book *Blink* (see Appendix C), Malcolm Gladwell advocates that we listen to that gut sense we all have about people and situations.

When you first started pondering whether to open your nest, did you have a gut feeling about it? Before you even begin to develop your list of pros and cons, listen to your instincts. Looking at the positives and negatives in a logical way could actually confuse your gut by adding extraneous data. Your intuitive feelings will likely be more accurate than any logical thinking would be!

Ground Rules

As you decide whether to open your nest, explore your gut feelings first. Your intuitive feelings may reveal more, and may produce more surprising, pertinent information than would a seemingly more logical consideration of the situation.

If your gut is quiet about what to do, however, or your spouse's gut is telling her something different than yours, then you should look at some pros and cons. Keep in mind your initial intuitive response, and then examine every possible angle, and you'll come to a decision that will be right for your family.

Examining Your Personal Perspective

In everyday life, we constantly make lists of pros and cons. We may think through the benefits and drawbacks of buying that double latte, or we may consider whether to confront a co-worker about her cell phone's annoying ring tone. We examine our thoughts and feelings about a subject in order to come to a decision.

When deciding whether to open your nest to your adult child, making a pro-con list can give you and your child a chance to look at a comprehensive overview of the costs and benefits of becoming roommates again. And it all starts with understanding how you uniquely perceive the situation before you.

We all view life through our own unique personal lens. We develop that lens through the things that influence us most, like our culture, our family, our life experiences, the communities we live in, and our religious beliefs. Each one of those factors affects how we look at and respond to a particular situation. When you and your adult child are considering the benefits and drawbacks of living together again, your personal lens will inevitably inform your thinking.

Your culture, or your beliefs and traditions, may influence you and your child's perception of living together. In your culture, for instance, is it expected for an adult child to remain at home until marriage (and possibly even after that), or does it value independence more than anything else?

Your family history matters, too, as you may mimic what previous generations have done—or even do the polar opposite and possibly overcompensate in the other direction! To help you become aware of any (potentially unconscious) biases you could have surrounding this issue, consider what your parents would have thought about you if you wanted or needed to move back home. Perhaps they allowed—or even expected—you to remain home until you married. Maybe your family of origin responded the opposite way, making you feel like you were being lazy or a failure if you needed to live at home while working your way up from dishwasher to manager.

> ### Under Your Roof
>
> Whether you are conscious of it or not, you may respond to your adult child just as your parents did to you. Are you judging your child negatively for returning home, as your parents might have judged you? Examine any biases you're aware of, and try to open your mind to the needs of this new generation.

Your own life experiences will also color your perception of possible open nesting. Maybe your stay at your mother and father's after college was a tumultuous one. Did you regress back into childhood and assume parent-child roles, or did you bond with your parents differently—as adults? Would you want your children to have the same experience?

Your lens is also influenced by how your friends, co-workers, neighbors, and family members feel about the open-nesting scenario. What would your current community think about an adult child living at home? What kinds of experiences have you, your friends, or your family members had with open nesting? For instance, would your friends and family envy your chance to spend more time with your son or daughter, or are they more likely to look down their noses and remind you that *their* daughter is a highly paid lawyer in Manhattan?

All these different points of view and influences come into play when looking at the benefits and drawbacks of having your child move back home. When making your pro-con list, your lens will provide the framework for how you view the situation and what things are important in your decision-making process.

Making a Pro-Con List

In this exercise, the parents, the adult child, and any siblings who will still be living at home, should brainstorm—individually—about the pros and cons of living together again. Follow these steps:

1. **Get set up.** Find a large piece of paper and split it into a "pro" side and a "con" side.

2. **Brainstorm.** True to the spirit of a brainstorm, write down *every* benefit and drawback you can think of without judging, changing, or even discussing each item until after you complete the list. You don't have to show the list to your child/parent at this point, so don't worry about censoring your thoughts.

3. **Refine.** Once you've recorded every pro and con you can think of, you may want to create a second draft of your list. This time, you can write more important pros and cons near the top of the list, and put less significant ones closer to the bottom. You can also remove any items you're having second thoughts about.

4. **Share.** Parents should discuss each item on their final lists with each other (and/or a close friend or family member). Parents should ask to see the lists created by siblings who would still be living at home when the adult child returns. Adult children can discuss their lists with a friend or other favorite sounding-board, or just think hard about each pro and con.

5. **Decide.** Rather than basing any decisions on the *number* of items on each side of your list, focus on getting a sense of which side of the list *feels* more compelling to you. If your "gut" was quiet before you did the exercise, perhaps it will be speaking to you more loudly now.

Parents' Pros and Cons

As a parent, you can certainly think of several good reasons to welcome your child home. A full house brings drama, joy, and excitement as everyone's lives intertwine. It also gives parents and children a chance to bond over conversations and shared experiences.

Benefits of Open Nesting

Looking at the situation through your personal lens, your list of pros will be unique to you, but here are some common benefits to consider:

- **Have continued connection with child.** When your son or daughter moved out, communication likely decreased. During this phase of life, it's normal for your child to relate more with his friends than with his folks. When your child moves back in, you'll have a chance to get back into his life, and have an opportunity to share in his young adulthood.

  But privacy and independence will be important to preserve, no matter how tempting your desire will be to be a part of your adult child's life. You can keep a safe distance and still grow the connection between you two.

- **Get your "job" back.** What better satisfaction to see your child grow? You raised him, you watched him make friends, and you saw him develop into an independent adult. If your son or daughter comes home, you get your old job back, a position you held for more than 18 years—and one you were quite good at!

Remember, however, that your child is not a "child" anymore. Your job description will very likely change. You'll need to decide which aspects of the job to keep and which ones to eliminate as no longer relevant. You may choose to cook some meals, pay a cell phone bill, or resume a role as advisor or sounding board (only if your child wants these things, of course!).

Even though you may take on some old parenting responsibilities, they should not significantly curtail activities or time with friends or your spouse just to be around more for your child.

♦ **Draw strength as a family unit.** Perhaps the greatest benefit of opening your nest to your son or daughter is that you can help and encourage each other. Not only can you support your child through a crisis, but now that your child is grown up, she may be willing and able to be there for you, too.

Keep in mind, however, that it's developmentally normal for your adult child to focus more on friendships than you at this stage. It's your son or daughter's job to become independent right now. If your child already feels satisfactorily independent, she may be entering the stage of becoming interdependent. Interdependence (discussed further in Chapter 3) is a developmental stage that is characterized by the ability to be there for family members without feeling pressured to do so. That's a sign of real maturity—and it's a real plus for you!

♦ **Ease financial burdens.** During these challenging financial times, families can come together and share expenses. Your child can help pay for groceries or the electric bill, and you can help her by charging minimal—if no—rent. If you decide to join financial forces, though, be sure to establish clear expectations for everyone.

♦ **Get some help around the house.** Your son or daughter can—and should—help with some chores, too. He could do some cleaning, cooking, pet care, or house sitting. He could do some yard work or help whittle down that "honey-do" list. Your child could be an extra set of hands to help around the homestead. Again, it's important for parents and adult children to form an agreement about expectations for chores before resuming living together.

Drawbacks to Open Nesting

While there are lots of benefits to welcoming your son or daughter home, there are plenty of drawbacks, too. You've been looking forward to spending time with your spouse, pursuing hobbies, or diving into the next phase of your life. In fact, parents with children who have left home report higher satisfaction with life because they're free to pursue their own dreams. A son or daughter coming home could change those plans if you feel you need to resume your "parent" role.

When creating your pro-con list, consider these drawbacks when deciding whether to allow your child back into the nest:

◆ **Hard to move on.** When your adult child moves back home, you may feel the need to put your life on hold. You may be reluctant to return to work or take that promotion. You may not have as much time for current friends, not to mention for making new friends. You may feel less able to focus on developing your relationship with your spouse. You may worry that all those empty-nest joys will need to be put on the back burner.

Though it may be difficult to move on, you should do it anyway! Don't let the return of your adult child cause you to regress to old behavior. Take that promotion. Develop those hobbies and friendships. Spend time with your spouse and younger children. Even these obstacles can be mitigated with some extra effort.

◆ **Your worrying returns.** Remember when your child went on her first date? Or when your son borrowed the car for the first time? That worry that kept you up all night is bound to return. It's natural for parents to worry about their son or daughter, and when she is back under your roof, you'll feel that extra responsibility again. Your worry may no longer be about your daughter drinking at an after-prom party, but you very well could toss and turn while wondering whether she will come home or spend the night at her boyfriend's house.

You may feel tempted to micromanage your adult child's life if you worry that she is making poor decisions. This can add stress to your life—and cause resentment in your child, because she will not feel as independent as she would like. You will need to trust that your child can make her own choices—and learn from her mistakes.

◆ **Stress from sharing space.** Did you turn your son or daughter's room into a den or a sewing room? Maybe you knocked down some walls and expanded the kitchen, or maybe you downsized and moved into a two-bedroom apartment. With your adult child coming home, you'll need to share your space with your child— and not just relinquishing his bedroom! You'll be sharing the living room, the kitchen, the garage, and the bathroom, and that will certainly create some tension in the household.

◆ **Triangulating child.** For the past 18-plus years, you and your spouse have focused your energy on raising your son or daughter, building your careers, and taking care of your home. With all of these outside focuses, you may not have noticed marital tension, but it surfaced as soon as your child left. When your son or daughter comes home, that strain could subside again—but not in a healthy way.

> **A New Generation**
>
> Triangles form in an emotional relationship to stabilize two people under stress. For example, a husband and wife enduring marital struggles may triangle a third person— an adult child—to divert anxiety in the relationship.

Putting your child in between you and your spouse to alleviate marital tension is called triangulation, and it prevents you from moving forward with your relationship. In the empty-nesting (or open-nesting) life stage, it's appropriate to come closer as a couple by working through relationship issues, and putting a child in between you may hinder that.

◆ **Growing resentment.** If your adult child starts to take advantage of your generosity, or you see her turn into a mooch, you may develop some resentment toward your child. You son may overstay his welcome or turn the living room into a pigsty, or your daughter may be spending her money going out with friends and not saving it. You could very quickly grow some negative feelings toward your child, which could turn into some unpleasant arguments.

♦ **Added costs.** You can expect additional costs associated with your child being home. They could be literal costs, like your grocery budget, or time and energy costs like increased worry or feeling like you need to be home rather than pursuing your passions.

Remember, you don't have to pay for your child's expenses or spend more time or energy on him. It just may be hard to avoid—especially for some parents. (And you know who you are!)

Pros and Cons for the Adult Child

For many adult children, the decision to move back home is easy. Who wouldn't want to return to the comforts of Mom and Dad's house, complete with a comfy bed, three square meals, and a 42-inch high-definition television and free Internet! It's a chance to get back on track in the safety—and among the niceties—of the family home.

Benefits to Moving Home

As your son or daughter puts on his or her personal lens and considers whether to move back home with Mom and Dad, here are some common benefits for your child—and you—to consider:

♦ **Financial support.** One of the most significant benefits of your adult child moving home is the financial support you provide by giving your child a roof over her head. While your daughter lives at home, she can save money to purchase a house, or your son can start paying off his student loans.

In addition to providing housing, you may choose to subsidize your adult child's grocery bills, car payments, cell phone, or other expenses. While your son or daughter may eat this up, consider that it may be hard for your child to learn to become financially independent if he takes too much advantage of your generosity. Paying some household bills or kicking in for groceries will help teach budgeting skills that the young adult can take with him.

- **Creature comforts.** Your adult child can enjoy nicer, more spacious surroundings than what he could afford on his own. Plus, with amenities like a stocked fridge, washer and dryer, and cushy couch, your child may never leave!

- **Emotional support**. Another tremendous benefit of moving back home is the emotional support a family can provide. There's no place like home for sympathy and guidance through a hard time, like job loss, a bad breakup, or a financial crisis. And when the child learns how to manage the situation with the family's help, she will become better able to handle similar problems in the future.

- **Welcome company**. Thanks to later marriage ages combined with a cultural emphasis on rugged individualism, never before have young adults spent so much time living alone. Solo-living (or living with a pretty independent roommate) can feel isolating compared to family life or college life. Moving back home can provide that personal interaction a young adult craves.

- **Stronger parent-child bond**. For many families, when a son or daughter leaves home, a natural separation occurs—usually in the form of the child's rebellion—that helps to launch him from adolescence into adulthood. In the launching process, teenagers typically go through a rebellious stage, followed by a stage called rapprochement, which is a better, more mature relationship between the parent and child. This brief bond is often followed by another rebellious stage, referred to as "spoiling the nest," when the child goes through an upsurge in provocative tendencies shortly before he leaves.

  When the son or daughter returns to the nest, that move may once again strengthen the adult child's relationship with his parents. He does return as an adult, and this enables a new bond to form, one in which everyone is an equal.

Drawbacks to Moving Home

To some young adults, having to move back home is anything but positive. A young man who isn't able to land his dream job is forced to live with Mom and Dad, or a young woman whose marriage disintegrated must go home to start over again. It marks a time when things aren't going right, and a last resort is to run to one's parents.

Though returning home surrounds the young adult with family and comfort, she has some significant drawbacks to consider. The topics listed below represent some common causes for concern:

- **Economically dependent, emotionally independent**. Your son or daughter will be relying on you for financial stability while trying to be an individual. Your child, for instance, may gladly accept the housing and other benefits you're providing, but will frown at your "parenting," like when you try to regulate her behavior. Your adult child will no longer think her 1 A.M. curfew (or any curfew at all) from high school days is acceptable, as it impinges on her highly prized adult independence.

- **Restrained freedom**. Let's be honest. Living at home is nothing like living on your own. Under Mom and Dad's roof, it's not likely that a young man could host the same kind of parties he would when he was on his own. A young woman may not be able to crank up the music like she would in her own apartment. And overnights with boyfriends or girlfriends may be a challenge—if not impossible, depending on the household. When an adult child moves back into his parent's home, he will need to exercise some restraint—which can be a lot to ask at this stage of life.

- **Parents "buying" influence**. This is where guilt comes into play. If Mom and Dad are providing free room and board, the adult child may feel compelled to essentially do what the parents want. The parents, for instance, may insist the child have a certain type of job while he's living at home, or try to influence which friends their adult child spends time with. They could also guilt-trip the adult child into caring for pets or younger siblings, doing lots of other chores, or being home for dinner every day.

- **More strain on relationship**. If the parent-child relationship was strained before the young adult moved home, it could only get worse. Old arguments may surface. Parents could fall back into disciplinarian roles while the child struggles against it. It can certainly cause stress in the household, and may do damage to the relationship. Some families simply get along better when everyone has her separate space, allowing everyone more autonomy. If you suspect your family falls into this category, you should trust your instinct and not assume that things will be different this time!

◆ **Privacy issues.** Parents may be tempted to go through their adult child's belongings when he's at work or school. They may log onto their daughter's MySpace page or Facebook wall and see who her friends are and what pictures she's posting. They may eavesdrop on telephone conversations. Even worse, Mom may burst into her son's room while he's entertaining—which could embarrass everyone involved! When a young adult moves back home, she'll need to get used to limited—or no—privacy.

◆ **Cultural, social embarrassment.** In some cultures and social situations, a young man may feel embarrassed about returning to the nest due to an emphasis on independence. His friends may mock him for "living in his mom's basement" or he won't be viewed as an adult by employers. It can weigh on his ego and confidence.

The Least You Need to Know

◆ When considering whether to invite your son or daughter back to the nest, go with your gut feelings first. Your intuition is likely to be more accurate than you think!

◆ Developing a pro-con list starts with learning your own personal lens, the way you view your world.

◆ Parents and adult children have many things to consider when deciding to live together again. Develop your pro-con list, brainstorm your ideas, refine them, and talk about them before you make your final choice.

◆ Once you and your adult child have made the decision to share a roof (again), be ready to formulate a plan that outlines how the new living arrangement will work.

Chapter 3

A New Way to Be Independent

In This Chapter

- Identifying healthy development goals for each family member
- Understanding critical goals for the family
- Being independent *and* interdependent
- Learning how to set and achieve your goals

George Bernard Shaw tackled the issue of independence in his famous play *Pygmalion*, where he wrote: "Independence? That's middle-class blasphemy. We are all dependent on one another, every soul of us on earth."

He had it right. Despite contemporary society's drive to produce independent, autonomous individuals, we depend on each other to communicate, collaborate, trust, and empathize. That sentiment is true for many families today, especially those families that allow an adult child to move back home when he needs the emotional or financial support of his parents.

Independence, then, takes on new meaning when a young adult moves back into the nest. Though each individual in the family will want to pursue her own dreams and aspirations, everyone will need to work—and live—together again, at least for the short term. Dependent on one another, they will be working as a family toward a larger, common goal, like helping Junior move into his own apartment.

It is possible for every family member to maintain independence and achieve his or her individual goals while everyone is living together. More important, a family may even foster a new, healthier way for the adult child to be independent, in which he can successfully launch on his own but also relate to the parent as an adult peer. And that's the ultimate goal, isn't it?

What's Good for the Individuals

In human development, people accomplish certain physical, intellectual, social, spiritual, and emotional life-cycle tasks as they progress through their lives. By "life-cycle tasks," we mean specific physical, intellectual, social, spiritual, and emotional milestones an individual hits as she progresses through life. Conscious of them or not, you could translate these tasks into "life goals," which take on real meaning when practiced in everyday life. Your goal of retiring by age 62, for instance, falls in line with a life-cycle task of planning to handle work transitions and retirement.

As each family member moves through his life cycle, he passes various developmental milestones. Each family member's movement through the life cycle affects—and is affected by—other family members and what their developmental needs are. In an open-nesting environment, your needs will most certainly interact with your son's or daughter's, not to mention your partner's and that of your younger child. So it's helpful to understand how individual goals typically relate to each other.

Ground Rules

As you read though this chapter, think carefully about your personal goals for this stage of your life. Consider how you can continue pursuing those goals while your adult child is living with you.

The Young Adults' Life Tasks

In early adulthood, from age 21 to 35, a young adult is developing the ability to participate in intense relationships committed to mutual growth, and participate in satisfying work. Your son or daughter is developing an independent self (which we discuss later in the chapter) who is able to interact with others in a meaningful way. Psychologically, your young adult child is entering and engaging with the world.

Family therapists identify specific life-cycle tasks or goals that a young adult should achieve in early adulthood. They're not set in stone; individuals certainly take different paths as they travel through life. They're meant to be general guidelines to help your child form goals that will ultimately enable her to launch from the nest.

The first of these milestones is an ability to differentiate herself from her family of origin, questioning how her beliefs, values, and habits differ from those of her parents. Your daughter, for instance, may choose to explore a different religion, or your son may choose to move from the country to the Big City.

A young adult should also show an increased ability to care for himself and his own financial, emotional, and spiritual needs. At this stage, your son or daughter should be employed or attending school, have a solid network of friends, and even start soul-searching a bit.

Another life-cycle task in early adulthood is your child's awareness and ability to deal with his own and others' sexuality. At this stage, the hormones are flowing, and your son or daughter should understand how to manage (and respect each others') natural biological urges.

Similarly, a young adult should also be developing an increased discipline for physical and intellectual work, sleep, sex, and social relationships. At age 20, for instance, your adult child has had enough life experience to form patterns of behavior, to understand the need for regular sleep, interaction with friends, and exercising her body and mind.

Goal-setting, which we delve into later in this chapter, is yet another life-cycle task for young adults. Your son or daughter should be honing the ability to create long-range life goals regarding a career path, intimate relationships, family, and community. He should also be showing more tolerance if those goals aren't immediately met—which may be challenging to some young people today who may be used to instant gratification!

Young adults should also be developing the ability to navigate through evolving relationships with parents, peers, children, community, and co-workers. For instance, if your daughter earns a promotion at work, she'll quickly begin to learn how to handle office dynamics (or politics, depending on the workplace!).

Nurturing is another task that both young women and men begin to develop in early adulthood. Your son or daughter should be learning how to support family and friends physically and emotionally. And if your adult children have children of their own, they'll be developing the ability to support those children (your grandchildren!) financially and emotionally.

In sum, the key developmental tasks for a young adult include forming mature, responsible, give-and-take relationships (or even taking on a care-giving role), and gaining the discipline to form and reach longer-term career goals. The young adult no longer shows the self-centeredness of adolescence or espouses the mentality of "instant grati-fication." (Even if you can't seem to get those iPod earphones dislodged for more than a 30-second conversation!)

The Parents' Life Tasks

Parents, too, will have developmental goals to fulfill. If you have a child who is in her early adulthood, you're likely to be between your mid-40s and your late 60s. You're accomplishing the tasks of middle adulthood and entering your wisdom years, when you reflect on your experience and focus on helping others and serving your community.

During this phase of your life, your main focus is nurturing and sup-porting your children and partner, as well as taking care of your older relatives. You're probably still working and bringing home a paycheck. You may be helping your mother pay her bills or taking your father to doctor appointments.

At this stage, you also may be looking at how satisfied you are with your career, and whether it's bringing in enough money to carry you and your family to retirement—and beyond. You may even go through the "mid-life crisis," where you dramatically change professions or take up an exciting new hobby.

In middle adulthood, you'll also be deepening your friendships, holding on to your tried-and-true cronies—your BFFs—and sharing more and more common experiences together. You may travel to Europe with your girlfriends, or go on that deep-sea fishing excursion with your buddies.

You may also begin to involve yourself more with your community, looking out beyond your own life and volunteering to improve your neighborhood park, help disadvantaged people, or even get more involved with your church or temple. As you begin to settle into retirement, you'll have more time to better the world around you.

Finally, you'll be solidifying your life and spiritual philosophies. With age comes wisdom, and at this stage, you should begin to have a pretty good sense of why you're here, what your life goals are, and what you believe in.

So during this stage of your life, you'll be really striving for a balance between self, family, and community. Your life tasks will involve looking at your own life and assessing its progress, taking care of your family, and seeking ways to better yourself through close friends and community outreach.

What's Good for the Family

When parents decide to allow their adult child back into the nest, it's critical that the process benefit the family as a whole. Ideally, the family will grow together into an interdependent unit, sharing in each other's lives while at the same time enabling members to maintain developmental goals as well as personal goals and aspirations.

Before interdependence can be achieved, however, young adults must establish their own independence (and you must maintain your *own* independence as well). Let's take a look at how a young adult finds autonomy.

Establishing Independence

Like generations before them, today's young adults exist in a time when individual expression is paramount. In this new century, however, they're branding themselves with tattoos and making statements with their clothing. They're embracing technology and communicating in their own IM and texting language. They're doing everything they can to differentiate themselves from the older generations.

Seeking that individuality is a normal part of development. During a child's early- to mid-teen years, a young man or woman works very hard to establish independence from parents. Granted, young people may be expressing themselves a bit more ... err ... creatively than they did 50 years ago, but it ultimately leads them to launch from the nest.

Ground Rules _____

Adult children returning to nest may still be in some stage of establishing their independence, so be ready for it. If for some reason—like parental over-control—your son or daughter wasn't already able to develop independence, you may find your child in a rebellious phase or dating a seedy character.

Teens and young adults assert independence in a number of ways. Some common behaviors include the following:

- ◆ A young woman may avoid her parents and/or act rudely to them. For instance, your daughter may come home from work or school and head right to her room or she may decide to text her friends during dinner. These "offensive" acts are actually very important developmentally—the child is essentially separating from Mom and Dad.

- ◆ A young man intentionally breaks rules. Mom may ask her son to call if he's coming home late and he doesn't do it. Dad may tell his son to keep his room clean and he refuses. They're bucking the establishment—you!

- ◆ A young woman dresses knowingly in a way that makes her parents uncomfortable or hangs out with peers who don't share their values. The young adult is distancing from her parents by doing things that they don't approve of.

This independence-seeking ultimately leads to rapprochement, which is the tendency for older teens, especially high school seniors, to act more easy and open with their parents. They've achieved a separate identity from them, and they're comfortable hanging out with Mom and Dad and sharing a more adult-like relationship with them.

When children leave the nest, they experiment even more with establishing their independence. They may run with the wrong crowd for a while, move far from home, or dabble in alternative lifestyles. Most will want to run their own lives, except when they need financial handouts from Mom and Dad, or need help with situations where they find themselves in over their heads, like an arrest, paying taxes, or an accident or illness.

Forming Healthy Connectedness

Once a child gains independence from parents, she begins to separate her own intellectual and emotional functioning from that of the family. It's a concept called differentiation or individuation.

Renowned psychiatrist Murray Bowen, a pioneer in the concept of differentiation, spoke of individuals functioning on a continuum of differentiation. Those with low differentiation depend on others for approval and acceptance. They're the types who can be emotionally manipulated or force others around them to conform to their way of thinking.

On the other end of the continuum, those who have a well-differentiated self, recognize that they need others, but they depend less on others' approval and acceptance. Rather than conform or force others to conform, they consider situations thoughtfully and come to their own conclusions about a course of action.

In a household with adult children, the key is for all family members—not just the child—to work toward a higher level of differentiation. This helps the family to solve problems and

Under Your Roof

If members of your family experience anxiety when someone else expresses a difference of opinion or makes a choice that they disagree with, it's likely that your family would benefit from becoming more differentiated.

create shared goals. When each family member has a clearly defined and differentiated sense of self, he is able to set emotions aside (which often run amok during family conflict), stay calm and clear-headed, and assess the situation using the facts at hand. Each family member acts in the best interest of the group and steers clear of reacting to relationship pressures.

Family members who are well differentiated can function successfully in an interdependent, or mutually dependent, household. This interdependent family, then, can begin to foster healthy connectedness, which means that each person in the household is separate emotionally and intellectually from the rest of the family, but each remains part of the family unit, sharing genuine thoughts and feelings with the other family members. It's an excellent goal for a family to aspire to.

Are We Differentiated Yet?

In an open-nesting situation, it's helpful to understand where your son or daughter is in terms of differentiation of self and whether she can have an interdependent relationship with you. A young adult who is differentiated enough to be interdependent exhibits these skills:

♦ A young man or woman is able to participate in cooperative activities at home, work, and school. Your son, for instance, is able to work together with a classmate on a school project, or your daughter is able to cooperate with a sibling to clean the house.

♦ A young adult is able to express a full range of emotions and tolerate them in others. For example, your son is able to empathize with his friend's breakup, or your daughter is able to accept responsibility for getting a traffic ticket.

♦ A young adult is able to express her differences of belief or opinion to others without attacking them or becoming defensive. She is able to articulate thoughts and concepts without letting emotion take over.

♦ Your daughter or son is able to relate with openness, curiosity, tolerance, and respect to people who are different. This young person, for instance, is able to set aside cultural or religious differences and befriend a new neighbor.

♦ He is able to nurture, care for, and mentor others, and will also accept help and mentoring. For example, a young woman is mature enough to serve on the community library board and to lead younger siblings at home in a family reading group.

Now, parents shouldn't necessarily assume that their adult child will be ready for interdependence. Your son or daughter may still have some growing up to do! So how do you actually help your child first feel independent from you? Try these techniques:

♦ Allow your son or daughter to express differences of opinion without shutting him down, or judging, criticizing, or punishing. Your son, for example, may choose to express himself with a new tattoo or outrageous hairstyle. Instead of telling him how silly or unprofessional he looks, accept his new 'do just as you would if a friend did the same thing.

♦ Become more tolerant of different values or preferences your child is exhibiting. Say your daughter decides to espouse a different political party. Rather than push your own views or coax her toward the other side of the aisle, ask her to explain her reasoning. Listen to her point of view. Strike up a healthy debate. Who knows, you may learn something, too!

♦ Let go of your desire to control your child's behavior or choices. For instance, if your son tends to leave a trail of clutter wherever he goes, rather than demand he clean up after himself—or else!— gently remind him of your cleanliness standards and request that he comply. If need be, pull out the contract (see Appendix A) he agreed to and signed.

Whether your adult child is ready for interdependence or needs more time to establish independence, it's important that the family's goal-making process continue to grow and nurture the entire family, rather than keeping parents or kids locked into roles of the past where relationships may be held back rather than progressing forward.

Treating Them Like Adults

When adult children move back to the nest, they may assume that they're completely independent from their parents. They've lived on their own, made their own decisions, worked through some mistakes—they're independent adults, all grown up … right? Who needs parents' old rules and expectations? Your son or daughter may test this theory by staying out all night or having a significant other stay over—things that were forbidden when they were in high school.

If you already increased your tolerance for your child's independence, but certain behaviors are still unacceptable to you, try using a relational approach with your son or daughter. Talk about your feelings and needs relating to the child's behavior rather than threatening consequences or punishments. Treating your son or daughter like an adult will help foster an interdependent relationship—which can lead to healthy connectedness.

Setting and Achieving Your Goals

Having looked at some typical overarching individual goals and family goals, let's get into what *your* specific goals are and how to actually accomplish those goals in an open-nest household. It's easy: all you'll have to do is set goals, share them with each other, and accept the fact that you may need to compromise if your goal interferes with your family's overall pursuits.

When you set a goal, you articulate a desired outcome and resolve to achieve it through attainable steps. For instance, a parent would say, "I will retire by April, one year from now." It forces you to state what you want to have happen rather than having a vague sense that you want something to change.

When you set goals for yourself or your family, the first step is to make the goal as specific as possible. This specificity helps you know if you are indeed reaching the goal. Instead of a mom saying, "When the kids are out of the house, I want to have a more active social life," she would outline what that actually means. She might say, "I will join the gardening club, volunteer once a week at the children's literacy organization, and play golf on Fridays with Susan."

To figure out how to work toward the goal, break the goal into smaller steps. This helps to keep you from getting stuck by not knowing what to do next (and helps you avoid procrastination). For instance, the mom in the previous paragraph would say, "To join the gardening club, I first need to look up online what the membership process entails. Then I need to contact a current member and ask her to nominate me. I then need to work the $500 membership fee into our family budget."

Challenges in Process

For some people, goal-setting can be incredibly motivating. It seems to spark a spirit of competition, even if the competition is with themselves. Many goals, however, are unrealistic and set up the person for failure and disappointment. For instance, if your goal is to save $15,000 in three months or lose three pounds a week, how will you do it? After looking at a goal more closely and outlining how you'll achieve it, you may realize it's unrealistic and decide to scale it back.

Under Your Roof

Don't set yourself up to fail. Many people give up on their goals because they set unrealistic expectations, vaguely articulate their aspirations, or find themselves stuck in the details of goal-setting. Make sure your goals are attainable, establish benchmarks, and stay focused on the prize. You'll gain a sense of achievement—and success!

Many goals are too vague, so goal-setters may not even know if they've reached their goals. "I'd like to spend more time with my family," the goal of many a hard-working dad, would technically be reached if he spent an extra five minutes with his family each week. This is probably not what he really means by "more time." It would be much better to make a more specific goal, like "I will have dinner with my family five nights a week."

For some people, breaking goals into steps or sub-goals can be time-consuming or confusing. They may find themselves stuck in the mire of list-making or outlining unclear lists. It can take some discipline (and maybe some research) to break goals up into achievable steps.

Your goals can also conflict with other people's goals—especially in an open-nesting situation. This can make it harder to commit to the goal in the first place, as well as stick to it. If a mom wants to spend more time outside the house expanding her social circle but her adult son wants her to cook for him each night, their goals are in conflict.

Goal-Setting in an Open Nest

When an adult son or daughter moves back home, it can be easy for family members to automatically settle into comfortable, even stereotypical, old roles, like Mom as family chef, or son as an oaf who doesn't lift a finger. But it's important to acknowledge that each person in that family is in a new part of his life cycle, as we discussed earlier. Mom, Dad, and Junior may have to choose very consciously to live differently and occupy new roles.

A New Generation

Goal: Helping Junior move out by August.

Challenge: Family members disagree on how to best reach the goal.

Junior's point of view: His parents should pay for everything so that he can save all his money for when he moves out.

Parents' point of view: Junior should pay for his own expenses and learn effective budgeting.

The solution: Negotiation, compromise, and reasoning through the problem, with all parties contributing ideas. The family needs to remember that they're now in a new life stage with an adult child. Negotiation usually creates more goodwill than using old strategies like parents "pulling rank" or the child playing the yelling/slamming/rebelling card. If the negotiation is unsuccessful, the person with the most power over the situation will usually get the final say.

It's important that family members don't lose their individuality or independence and start to become resentful. This not only applies to the mom who wants to trade in her spatula for a more active social life, but also to the son who strives for more independence—he no longer wants to be obligated to attend church with his parents or to come home by midnight.

Family members should share their personal goals with each other and any conflicts should be discussed and resolved. Mom may need to tell son that her new goals for herself will not allow her to cook for him every night the way she did when he was in high school. She could potentially help him problem-solve how to reach his goal (of having a nourishing meal each night) without compromising what she wants.

The family may also want to band together to develop shared family goals like, "Let's organize the extended family to take Grandpa on a cruise to Alaska for his seventy-fifth birthday." Each family member could have specific tasks (sub-goals) to move the process along.

Goal-Setting Exercise

To get you started setting goals for yourself and your family, try this exercise. First, pull out a notepad, diary, or even your laptop—but use something that will have some permanence, as you'll be going back to it again and again. Next, write your goal at the top of the page, being as specific as possible. Below that, detail realistic efforts you'll take to reach the goal.

Here's an example for you to follow, just until you get the hang of it.

Goals for improving or enhancing my marriage:

♦ Go out on a "date night" once a week.

Steps for reaching goal:

1. Agree on a night with my partner.

2. Collaboratively put together a list of date ideas.

3. Decide who will set up the date each week.

♦ Go in for a "check-up" with a marriage/family counselor.

Steps for reaching goal:

1. Get partner to agree to attend.

2. Ask friends for names of good counselors.

3. Call counselor and set up appointment.

4. Jot down list of what I'd like to work on.

Goal-setting in the context of open nesting can truly enable each family member to work toward his or her own dreams and aspirations independently. The family may also choose to create family goals, working together toward a common purpose and fostering that inter-dependence that leads to healthy connectedness. The entire process can nurture and grow the family as a whole. The end result can be a closer bond between family members—and hopefully a healthy re-launch from the nest.

The Least You Need to Know

◆ As you progress through life, you accomplish certain physical, intellectual, social, spiritual, and emotional life-cycle tasks.

◆ A significant life-cycle task for your son or daughter is establishing independence. Once your child does so, the family can enter into an interdependent relationship, which is appropriate in an open-nesting environment.

◆ In an open-nesting environment, the life-cycle tasks of you and your adult child will inevitably intersect—and interact—so you should set both personal and family goals to ensure that everyone in the family is moving in a healthy direction.

◆ Goal-setting involves articulating a desired outcome and resolve to achieve it through specific, attainable steps.

Chapter 4

We Are Family

In This Chapter

- ◆ Looking at traditional family structures
- ◆ Understanding more modern family structures
- ◆ Examining how families function in a healthy way
- ◆ Exploring family dynamics in the open-nesting environment

The makeup of the American family is more fluid and diverse than at any other time in our nation's history. From traditional families to same-sex partners with children to single parents by choice, just about anything goes. No matter its makeup, however, the concept of family remains as a group of people who function as a unit, its members drawing strength and security from each other.

For most families, as it probably is for yours, having an adult child move back home is a one-time, temporary event. The move is most successful when family members define clear boundaries, understand what to expect, and function together in a healthy way. This is especially true if there are younger children still at home who'd moved out from under the older sibling's shadow only to discover … big sister or brother is back!

If you had to define your family and rate how it functions, how would you do it? Perhaps you're single or divorced with an adult child—who you consider your BFF. Maybe you're remarried, and you and your spouse have formed a blended family, complete with several adult or nearly adult children who are fiercely independent. Your family's dynamics may influence your decision to open your nest to your adult son or daughter.

In this chapter, we'll explore more in-depth strategies for understanding what family means to *your* family—and whether it's a family that can succeed at open nesting.

The "Traditional" Family

When many people picture the "traditional" family, they bring to mind an *Ozzie and Harriet* or *Leave It to Beaver*-type archetype. There's a husband, the breadwinner and head of household; wife, the stay-at-home mom and loving supporter; and two children, who often find themselves embroiled in relatively harmless shenanigans. Everyone gets along, everyone knows his or her place, and the family functions in a healthy, positive way.

Today's typical American family looks nothing like the Nelsons or the Cleavers of the 1950s and 1960s. Families come in many forms: multigenerational extended families of three or four generations, gay or lesbian couples with children, remarried families with shifting numbers of children who belong to several households, single-parent families, and unmarried couples with children, just to name a few.

Under Your Roof

Despite the reality—and acceptance—of today's varying family structures, many people still consider a "traditional family" to be that 1950s nuclear-family throwback: a heterosexual, legally married couple and their 2.5 children. This family form is all too often considered the ideal manifestation of "family," but as our changing culture has shown us, we may want to rethink that ideal—especially as the look of families continues to evolve.

But regardless of its makeup, most families in the United States do tend to be nuclear families, indeed comprised of two adults and children (either their own or from previous marriages). The Webster's definition, "A basic social unit consisting of parents and their dependent children living in one household," can be expanded to include unmarried couples with children and same-sex households with children.

Modern Configurations

Many modern families, though, fall outside the realm of "parents and their children." In an open-nesting environment, these variations can spawn their own sets of challenges that you should be aware of before you decide to open your nest to your daughter or son.

Single/Divorced Parent

There are more single-parent households (either by choice or from divorce or death) than ever before. According to the U.S. Census Bureau, 9 percent of households were headed by single parents in 2006, up from 5 percent in 1970. And according to Families and Living Arrangements, in 2006 there were 12.9 million one-parent families, 10.4 million single-mother families, and 2.5 million single-father families.

If you're a single mom or dad with an adult child, you may face some unique circumstances if you allow your son or daughter to move back home:

♦ **Less privacy when dating.** Your child will know all about your "overnights," which could be embarrassing for everyone involved! Your daughter may overhear your phone calls. Your son may accidentally read a private e-mail or text message. You and your child may be dating at the same time—in which case, you may need to communicate and coordinate about overnight guests, or simply about who will be home when. If your adult child moves home, you'll both need to pay particular attention to privacy and personal space.

♦ **Your child may not like to see you dating.** If you're recently divorced or your spouse passed away, your child may not be ready to see you dating—not to mention being intimate with—another person ... even if it seems that child should be old enough to have the maturity to handle it.

◆ **Your child becomes your best friend.** If your daughter or son moves back home, you may rely more on the child for company than on friends or partners. It's easy to do. You may be tempted to help your son with his love life or go shopping with your daughter, creating shared experiences and becoming closer as peers.

If you feel you should delay dating or spend less time going out with your friends because of your son's or daughter's presence, don't! Just as we discussed in Chapter 3, you and your adult child should be striving for interdependence, where you each exist as individuals but still remain connected to each other. Besides, you may start to resent the delay and put strain on your relationship.

Remarried With or Without Kids at Home

When new families come together, new dynamics come into play. If you're a parent in a blended family and your adult son or daughter comes home, consider whether your spouse and your step-children may see your child's return in any of the following ways:

◆ **Your adult child is seen as an unwelcome intruder.** Your child's homecoming may be seen as an unwelcome event by your spouse and/or the younger siblings or step-siblings. Family members may feel that they need to compete with your adult child for space, time, or money. A younger step-brother may need to give up his room for your son. A step-sister may feel like her one-on-one time with Mom will be set aside. You'll want to pay particular attention to siblings and step-siblings if you allow your adult child to move back home.

◆ **The move is seen as a bad investment.** Step-parents may feel competitive with their step-children for attention and resources because they don't have the biological bond that tells them to invest in that child. Your spouse may not feel comfortable footing the bill for his step-son's cell phone or his step-daughter's car insurance. Be sure to talk to your spouse and articulate any concerns you may have.

◆ **The whole family is happy to have your adult child back at home.** Some younger siblings—step or not—may be excited to have an older sibling back. She can give younger children rides to school or soccer practice, help with homework, or offer advice. It can be healthy for young people to have another "adult" presence around. Your spouse may enjoy having the chance to bond more closely with a returning adult step-child.

A New Generation

Because of the high divorce rate—at just about 50 percent, according to the U.S. Bureau of Labor and Statistics—and rate of remarriage, many families today look a lot like the Brady Bunch. According to the National Center for Health Statistics (2002), 54 percent of divorced women in the United States will remarry within 5 years; 75 percent of them remarry within 10 years.

If you and your spouse bring your own sets of kids to a marriage, and you're considering letting one of your older kids move back home, take the time to discuss the situation with *all* of your children. Reassure them that just because their older sister or brother is moving back, their own lives and routines shouldn't change.

Multigenerational Household

According to the U.S. Census, there are lots of multigenerational households, too, where Grandma or Grandpa live with adult children during their senior years, when more care-giving is required, or when Junior and his family live at home with Mom and Dad. About 5.7 million children, or 8 percent of all U.S. children, lived in a household that included a grandparent in 2006. The majority of these children (3.7 million) lived in the grandparent's home, and of these, about 60 percent also had a parent present.

A less-common social unit, these extended families are made up of a nuclear family and its immediate family members, including grandparents, aunts and uncles, nieces and nephews, and cousins. Extended families can also refer to spouses of children (remember Gloria and Meathead of *All in the Family* fame?)—so if your son or daughter moves home and brings a spouse along, you've got yourself an extended family unit. (Flip to Chapter 7 for more about this living arrangement.)

A multigenerational family is likely to consider it "normal" for an adult child to live at home. Family members may not have as many biases against open nesting … the more, the merrier! Unmarried adult children in multi-generational households, though, would probably find it even harder to have a partner spend the night or enjoy privacy within the house.

Understanding How a Family Functions

Whether your family is a "traditional" or a "nontraditional" one, the way it functions—and whether that functioning is healthy—is important to diagnose if you're considering opening your nest to your adult child. Families that diagnose the health of their relationship within the family may have a better chance at making the open-nesting experience a positive one.

David Olson, Ph.D., renowned family therapist and co-author with Candyce S. Russell and Douglas H. Sprenkle of *Circumplex Model: Systemic Assessment and Treatment of Families*, developed a method for describing and analyzing families and how family members interact with one another. The model assesses family functioning on the dimensions of cohesion, flexibility, and communication.

Families that function in an optimal way can withstand the anxieties that life brings—even having your adult child move back home. But families that fall out of the optimal range may be a bit out of whack—and are likely to land in Lauren's office when times get tough. Understanding where your family is should give you some guidance in determining how well open nesting will work for you.

Cohesion

The first dimension of family functioning, cohesion refers to the emotional attachment that family members have toward one another. Every family—from the traditional family to the nontraditional—has some level of connectedness. You can be attached to your family in lots of different ways, like sharing friendships, making decisions, sharing interests, having fun together, and spending time together. You share friends with your spouse. You go fishing with your daughter. You help your son through a breakup. When you think about it, you're attached in many ways with your family members.

Olson lists four distinct levels of cohesion, or attachment. Picture these on a continuum, with the first being overly attached and the last being overly independent:

♦ **Enmeshed** is where there is too little independence and differentiation among family members. Each individual heavily depends on one another and emotionally reacts to one another. Often, the family members have fewer outside friends and interests. Enmeshment of family conflicts with a child's developmental need to become independent and develop intimate peer relationships. Open nesting could be hardest for this group.

♦ A family with a **connected** relationship still has a high level of emotional closeness and loyalty to one another. Time together is more important than time alone; there is an emphasis on togetherness. Family members share friends, but they also have their own friends. Shared interests are common, as are some separate activities. Open nesting is likely to be successful for this group.

♦ A family with a **separated** relationship has some emotional separateness, but it's not as extreme as a disengaged family (below). Time apart is important, but there is some time together, some joint decision-making and support among family members. Activities and interests are generally separate, but a few are shared. Open nesting is also likely to be successful for this group.

♦ **Disengaged**, on the other end of the spectrum from enmeshment, is where there is limited attachment and engagement. Family members are unable to turn to each other for support. Open nesting may not really be a huge challenge for disengaged families, as everyone will tend to "do their own thing," which is more or less developmentally appropriate at this stage, anyway.

Both extremes—very high levels of cohesion (enmeshment) and very low levels of cohesion (disengagement)—might be a problem for individual and relationship development in the long run. People in connected and separated relationships are able to balance being alone versus together in a more functional way. Similarly, in terms of open nesting, connected and separated relationships are ideal, even leaning more toward the disengaged end of the continuum.

Flexibility

The second of Olson's dimensions of family functioning, flexibility refers to how much change can take place in the family. More behavior-focused, a family's flexibility involves how flexible or tolerant its leadership is, how its members negotiate, whether family roles can be changed, and what kinds of rules it has.

To function in a healthy way, families need both stability and the ability to change when appropriate. As with the cohesion dimension, the more balance a family has, the better.

Ground Rules _____

It's never too late to improve your family's dynamics, because your family is always evolving as each person responds to one another. Never underestimate the ripple effect one person's behavior can create in the family environment! Even the simplest changes can have important positive consequences.

Olson identifies four levels of flexibility. Just like the cohesion model, picture these on a continuum, with the first being anything but flexible and the last being downright floppy:

♦ **Rigid** means low flexibility. A rigid family has one controlling individual in charge. There is no negotiating about anything. Rules are defined and roles don't change. Rigid families will probably have difficulty with open nesting because roles will *need* to change when the adult child moves back home. The child will need to be treated as an adult to meet her developmental goals of autonomy and differentiating from family. Open nesting may also be harder on a controlling individual who will be doing more work trying to keep the adult child "in line."

♦ With low to moderate flexibility, a **structured** relationship tends to have a somewhat democratic leadership with some negotiations that include the children. Roles are stable and they're sometimes shared. Rules are firmly enforced, and there are few rule changes. Open nesting could prove to be challenging for families with this level of flexibility. To be successful, they will need to remain conscious of allowing family roles to change and rules to adjust (or be lifted) to suit the greater maturity and the life-cycle needs of the adult child.

♦ With moderate to high flexibility, a **flexible** relationship has an egalitarian leadership with a democratic approach to decision-making. Negotiations are open and actively include the children. Roles are shared and there is fluid change when necessary. Rules can be changed and are age appropriate. An open-nesting arrangement should be the easiest for this group.

♦ On the other end of the spectrum from rigid, **chaotic** is where there is erratic, limited leadership. Decisions are impulsive. Roles are unclear and they shift between individuals. This type of family would also have trouble with open nesting because expectations would not be clearly defined for each member, and roles within the family would be ambiguous.

Just like with cohesion, families that are rigid and chaotic could face problems with individual and relationship development. Structured and flexible relationships enable families to bend with the tide just enough. Rules are in place, roles are somewhat defined, and the family is able to negotiate through differences of opinion.

Communication

In any relationship, communication skills are paramount. This third aspect of Olson's Circumplex Model focuses on the family's ability to communicate with one another. Effective communication is key in an open-nesting situation, as the household is now made up of adults functioning interdependently.

According to Olson, some good communication skills include the following:

♦ **Listening.** Good listening skills include being empathetic to what another family member is saying and listening attentively. When you speak with your son, for instance, you should put yourself in his shoes, keep eye contact with him, and actively engage in what he's saying. While listening, a person must focus completely on what the speaker is saying, rather than planning a response. Certain nonverbal behaviors can help to show that you're listening. Nodding, maintaining eye contact, and saying "uh huh" tell the speaker you're hearing what's being said.

♦ **Speaking.** Good speaking skills include speaking for yourself and not someone else. You should talk about your own experiences, thoughts, and feelings. "I" statements can be helpful in communicating your feelings to others. For example, "I felt anxious when you didn't come home when you said you would." Also try to avoid reading the minds of other family members. Ask a family member to clarify something said or done, rather than trying to guess at motives. False assumptions can cause resentment all around. And don't be afraid to apologize for something you did that hurt another family member, even if it was unintended.

♦ **Self-disclosure.** A communication technique that helps to build an emotional connection with others is to openly share personal experiences or feelings. For instance, when talking with your daughter, it could strengthen your bond if you told her about the difficulty *you* had in finding your first job and making ends meet.

♦ **Clarity.** Communication involves articulating what you're saying in clear, concise, understandable terms. Sometimes, your audience—a young adult—won't understand esoteric references to obscure lyrics by The Beatles, so you'll need to be clear with what you're saying in terms your child will get. And as you can undoubtedly remember from your child's teen years, concise requests are much more effective than lengthy lectures.

♦ **Continuity tracking.** Staying on topic is another important communication skill. If your daughter is talking about her new iPod and all the new songs she just downloaded, don't jump in with your opinion about how all these new gadgets are destroying the social fabric of America. Stick with the topic she brought up. Sharing her excitement will certainly strengthen your bond.

◆ **Respect and regard.** Good communicators also show respect and regard to others. They avoid belittling, criticizing, or showing contempt, even when they strongly disagree. For example, if your son brings home a conservative (or liberal) friend, try not to get into an overly heated political discussion.

Family members will need to communicate their feelings and needs to each other in an open-nesting environment, so when assessing how successful your family might be in this type of arrangement, consider whether everyone's communication skills are up to par.

Family Dynamics and the Open Nest

Functioning as a family in an open-nesting environment may present some unique challenges for parents, the adult child, and any siblings still living at home. Even if family members are interdependent, remain flexible and communicate well, certain dynamics between family members may come into play. Here are a few common issues that may surface.

Feeling Like a Third Wheel

During this stage of their lives, parents should be coming together as a couple and forming a dyad, which is when two individual units—you and your spouse—come together as a linked pair, forming a unified front when making family decisions or confronting a family member about a conflict. An adult child moving back home may throw a wrench in that process. He or she may come between the parents (usually unintentionally), preventing them from getting closer and spending time together.

In some families, a coalition forms between a parent and a child, in which the emotional intimacy that should have been present between the spouses is replaced by a bond with the daughter or son. The "other" parent feels alienated (just as the child feels "parentified"). When this happens, it is usually a symptom of a marriage that needs help. If there is a risk of a parent-child coalition forming in your family, it probably already would have happened while the child was living at home the last time. If this is a concern of yours, the best course of action is to pursue professional marriage counseling.

To help you strengthen your union and come together while your son or daughter is living at home, try these techniques:

- **Establish an "executive system."** Studied in depth by family therapist Salvador Minuchin, an executive system is formed when parents make decisions as a couple, and then present those decisions to the child from a unified front. For instance, if you and your spouse are deciding what chores to assign your son, first discuss the subject together, come to a decision, and present your list of duties as a team. If your son then comes to you to negotiate, you can use a relational approach with him (see Chapter 3), sharing the decision that you and your spouse made.

- **Continue working toward life tasks.** As we discussed in Chapter 3, you and your spouse have your own sets of life goals to be working toward—including solidifying friendships and coming together as a couple. When your adult child moves back home, keep moving forward! Go on your weekly date night. Spend time with your friends. Continue volunteering for the literacy center. You should feel no obligation to stay home and cook dinner or change your routine in any way.

Younger Siblings Displaced

In a household with younger siblings, the return of the older brother or sister will certainly cause a stir. When the older sibling moved out, the younger sibling's position in the family changed from being the younger child to the *only* child (or the oldest child, if there are multiple siblings still living at home)—and now it's changing *again*.

The younger child may feel that she will no longer have the same access to Mom and Dad, and will have to compete for resources and attention. A younger boy who started playing soccer, for instance, may worry that his parents will pull him from the team due to a lack of funding because an older sister moved home. Or a younger sister may need to share the computer with her older brother again. Younger siblings will likely feel less secure about themselves and their places in the family hierarchy ... even wondering whether what they perceive as hard-won gains in family status will disappear or significantly erode.

To soften the impact on younger siblings, parents should be especially cognizant of what younger sons or daughters will have to sacrifice for the older child. Parents should try to keep everything as "normal" as possible, allowing the younger son to continue playing soccer or scheduling time for the younger daughter to have access to the family computer. Parents should make sure the younger siblings enjoy the same benefits they did before the adult child moved home.

Despite the change in the household, and a necessary transition period, younger siblings may actually enjoy having big brother or sister around. Imagine: a teenage girl can spend hours talking to her big sister about clothes, relationships, and careers, or a teenage boy gets rides to school or baseball practice in his brother's cool convertible. It's an opportunity for your children to share a more equitable status with each other, and to encourage an open acknowledgment of how much more mature each family member is becoming.

A Return to Old Roles

It's easy to fall back into old patterns and habits. When an adult child moves back home, he may return to the role of child (no matter what age!), and with that comes a feeling of powerlessness and of being dependent on Mom and Dad. But as we learned in Chapter 3, the adult child has his *own* set of life tasks to accomplish, which include developing into an independent, responsible, self-sufficient adult who makes good decisions.

If your daughter or son seems to be regressing, share these points to cajole your adult child back to an appropriate role:

◆ **Redefine roles in the family.** When your child left home the first time, she was still considered a child. Now, upon returning, you consider her an adult. Adults treat adults as ... well ... adults!

◆ **Embrace authority.** The adult child should remember that she has more authority and decision-making ability as an adult. A young woman, for instance, can decide how to spend or save her money without being "told" by her parents.

◆ **Take on more responsibility.** Because the child is now considered an adult by the family, he should be willing to take on more responsibility, too, like helping around the house, contributing money for bills, and generally acting like a grown-up!

While living at home, young men and women should continue to strive for independence and work toward pursuing their personal goals and developmental goals. Just because an adult child lives at home doesn't confer permission to kick back!

The Least You Need to Know

◆ Families today come in unlimited configurations, from traditional dad-mom-child units to more contemporary family models made up of a single or unmarried parents and children.

◆ Families function on three different levels, according to Olson's Circumplex model: its cohesion, or how connected it is; its flexibility, or how well it adapts to change; and its communication, or how well family members communicate to each other.

◆ It's critical that each family member focus on individual life tasks rather than return to old roles.

◆ Pay close attention to younger children still at home, as they may feel displaced when their brother or sister returns.

Part 2

Living Together

So now that you're all back living together as a family, what can you expect from one another? In this part, we'll take a look at how a household of adults (and maybe a few younger siblings) can live together with new, albeit shared, purpose. We'll see how you can creatively re-create the family to reflect the fact that everyone is all grown up now, thank you very much! We'll investigate physical boundaries, sleepovers, married partners (and dating, too), rent, chores, and responsibility.

Chapter 5

Stay Awhile ... Stay Forever!

In This Chapter

◆ What to do when your goals and your child's aren't aligned

◆ Making sure you and your partner share the same goals

◆ Your child's not prepared or equipped to leave—now what?

◆ What if you don't want your son or daughter to leave?

The decision is made: you and your partner have agreed to let your adult daughter or son move back home. You listened to your child's reasons for returning, you examined your family dynamics, and you went with your gut instinct that whispered, okay, this situation will work—*but for how long?* Did you really consider the practical and emotional issues of timing?

The length of time your adult child will need to spend back in the nest will vary depending on your child's unique situation. The amount of time you allow your child to stay will vary, too. When you agreed to let your young adult move back home, you likely discussed a schedule and formulated some loosely structured goals. Sometimes, however, life throws curveballs, and those schedules and goals you're holding onto fall by the wayside.

In this chapter, we'll look at some of these curveballs that may come your way and suggest strategies for handling them in a way that will benefit the entire family. These specific situations that we'll explore together commonly surface in open-nesting environments, and it's helpful to understand productive ways to manage them.

Mismatched Parent-Child Goals

This first situation centers on individual goals not matching up and creating conflict in the home.

Eric, 23 years old, moved in with his parents after graduating from college with a Bachelor's degree in art history. He lived in his parents' spare bedroom and worked as an assistant manager for a local pizza joint to earn some extra cash.

Eric's parents, both of whom work full-time jobs, welcomed their son back home. They understood that he could use some assistance until he found a higher-paying job. The father had lived with his folks after college, too, so he wanted to give his son the same support his parents showed him. The mother enjoyed "mothering" her son, so she had no problem with him returning to the nest. They both looked forward to getting to know their son as a young adult.

> **Ground Rules**
>
> When you're setting individual goals or goals with your spouse or partner, make them as specific as possible and outline clear steps for achieving desired outcomes.

When Eric moved back home, the family briefly discussed how long he would be living with Mom and Dad. The lighthearted conversation concluded with a loose "we'll see how it goes" time frame—but the parents had "three months, max" in mind.

Five months passed, and Eric—who thought he'd be able to stay a year or more—hadn't budged. He spent his time online and hanging out with his friends when he wasn't working. He didn't save money for his own apartment. He essentially returned to his childhood role and developed no plans for the future. But his parents grew impatient with Eric's seeming lack of progress, and wanted him to move out.

The Analysis

In this case, the child wanted to stay home longer than his parents were willing to allow. His mom and dad saw that their son needed help and they gladly opened their home to him. But in their minds, he'd overstayed his welcome. Eric should have moved out two months ago—or at least made more progress toward doing so.

The situation came about for a number of reasons. The family failed to outline a clear time frame for when the son should move out. The parents neglected to share their three-month goal with their son. The young man created no goals to establish his independence, so he took advantage of Mom and Dad's generosity and decided to take what he perceived to be a well-deserved break before pounding the pavement to look for a job in his field.

The Solutions

If you're dealing with an issue like this in your open nest, you should first look at the reasons why you want your child to move out in the first place. Listen to your gut instincts, as we discussed in Chapter 2. Perhaps you're worried that your child isn't taking the hard work of establishing a career seriously. Maybe you're planning to retire soon and you want to move on with your lives. Or you could just be tired of being a parent, of policing, disciplining, and staying up worrying at night.

Once you identify those reasons and instincts, see if your worries can be mitigated at all using these strategies:

◆ **Create a contract.** A written contract will go a long way to help parents and adult children clarify their goals for the length of the child's stay. (You can write your own, or you can flip to Appendixes A and B for sample contracts.) A contract makes the arrangement official and gives you some backup if your child overstays your warm welcome. A contract is also useful because it enables you to outline—and agree upon—conditions for your child's stay (like whether he'll be required to have a job, whether she'll attend school, or even whether rent will be paid).

◆ **Draft a plan.** If you worry that your young adult isn't making much progress or is even regressing to a childhood role, you can work together to develop a timeline with specific milestones. For instance, by March 1 your child will have saved $2,000 toward rent for an apartment. Setting goals like this can help keep ambitions on track, and it can help you encourage progress in a loving way.

◆ **Find a compromise.** Your family can also look for ways to give and take. In our example, if Mom and Dad want Eric to move out, but he's not making enough at the pizza place to fend for himself, they can compromise and extend his stay another two months until he lands his dream job working for a gallery curator.

Under Your Roof

Though you may feel like you don't need to draw up a contract or formulate a formal plan, do make it a point to sit down with your partner and adult child, discuss everyone's expectations (don't forget younger siblings who may still be living at home, too!), and write down your goals as individuals and as a family. You may be surprised what kinds of issues surface.

It's important for parents to remember: *you have the upper hand* when it comes to how long your adult child stays in your home. You should feel empowered to tell your child how long the nest will stay open—and when your young adult should fly the nest and leave home. It's your house, and even though this is your child, you're all adults now … and adults respect agreements and strive to honor them.

Conflicting Goals Between Partners

The second situation deals with partners who disagree about how long their child should stay.

The parents of Jen, a recently divorced 28-year-old woman, allowed her to move back into her old room while she started the arduous process of rebuilding her life. Because the breakup was so traumatic, the family wanted to "go easy" on Jen, and they failed to discuss any sort of time frame at all. They spirited Jen home to help her through this tough time without thinking it through.

Six months passed, and Jen's mother started to feel a little anxious about when her daughter would finally move on. She tried to share her concerns with her husband, who just said he felt Jen could stay as long as she needed to. So just as the family didn't discuss the situation together, Jen's parents didn't talk about it, either. Jen sensed the tension between her parents and felt something was up—like her welcome, maybe?

The Analysis

In this case, Mom and Dad had different goals and expectations for their daughter's length of stay. Mom thought that six months was enough: it was time for Jen to head out into the world again and get on with life. She felt it would be more developmentally appropriate for Jen to begin spending more time with friends and less with family. Plus, Jen's mom was ready to focus more on her relationship with her husband—without a third party present. Dad, however, wanted to protect his little girl and help her as long as she needed him. He wanted to be sure his daughter was emotionally ready to handle being independent again.

Before Jen moved in, her parents didn't talk about how long they would allow her to live with them or share their personal goals with each other. When her mother was ready for Jen to leave, she assumed her husband felt the same way. Jen's father, however, had different ideas in mind. Their goals conflicted because they didn't take the time to swap their desired outcomes.

The Solutions

Does this scenario sound familiar? If so, remember: because this time of your lives should involve coming together as a couple rather than splitting apart, it's critical for you to sit down with your partner and discuss your reasons for wanting a particular outcome. (We offer some communication techniques in Chapter 11.) Why does Mom want her daughter to move out at the six-month mark? Why is Dad fine with his daughter living at home indefinitely?

Once you identify your own lines of reasoning, you can then start a dialogue and begin to compromise, find a middle ground, and develop a plan together as a married couple. For instance, you may come to agree that you'll allow your daughter to stay for one year. You can then present your vision to your child, and then as a family you can develop a timeline with milestones so she can reach those goals with your help.

Ground Rules

Although parents should strive to compromise with one another, if one partner is adamant on a particular point, that partner should really be listened to. And if possible, the wish should be granted.

A couple should be coming together at this time in their lives, and they shouldn't let an adult child's needs come between them in their marriage, especially if the child moving back in would make one of them feel resentful or powerless.

Your Child's Not Ready to Leave

This third case focuses on adult children who lose their forward momentum.

Michael, 21 years old, moved back home to take a semester off from school where he was studying forestry. Before he moved home, he talked to his parents at length about his plans to take a break, regroup financially, and return to school in the fall.

Though their gut instincts told them otherwise, Michael's parents agreed to the arrangement, but they requested that he sign a contract and provide a written action plan with goals and milestones. The couple had recently retired and was hoping to do some traveling in the next year, so they remained hopeful that Michael would follow through on his promises.

When the end of July approached, Michael hadn't taken any steps toward leaving despite the plan, the signed contract, and the time frame with specific goals. He was like a deer caught in the headlights, frozen in time and not sure what to do next. He wasn't ready to leave.

The Analysis

In this scenario, Michael returned home because he was struggling with financial problems. He had every intention of returning to school in the fall—he had a plan in place, after all—but he failed to develop the skills to achieve financial independence while living with Mom and Dad. In addition, Michael was experiencing paralyzing anxiety that kept him from making any movement toward his ultimate goal of going back to school in the fall.

Michael suffered from generalized anxiety disorder. The American Psychiatric Association (APA) recognizes this condition when someone suffers from anxiety more days than not for at least six months, and the person finds it difficult to stop worrying. The APA says that three or more of the following symptoms tend to be present as well:

- ◆ Restlessness or feeling keyed up or on edge
- ◆ Being easily fatigued
- ◆ Difficulty concentrating or mind going blank
- ◆ Irritability
- ◆ Muscle tension
- ◆ Sleep disturbance, such as difficulty falling or staying asleep, or restless unsatisfying sleep

The worry will also cause significant distress or negatively impact a person's social and occupational abilities. Michael ended up in his predicament because he didn't have the tools to cope with his anxiety, and because he lacked some important life skills. Maybe he didn't know how to create a good resumé, or how to interview for a job. In addition, perhaps he had never learned the skills to budget his money and pay down his debt.

The Solutions

Though this example was fairly specific, the overarching idea includes many different scenarios. Sometimes, when an adult child like Michael moves back home, he stops taking the steps necessary to launch from the nest. Wondering whether he is ready and able to leave can compound his anxiety and lead to inertia.

Another common mental health problem to look for in your child is depression. When someone is depressed, five or more of the following symptoms are usually present:

♦ Depressed mood most of the day, nearly every day

♦ Markedly diminished interest or pleasure in any activity

♦ Significant weight loss (unintentionally), weight gain, or a change in appetite

♦ Not sleeping at all or sleeping too much nearly every day

♦ Appearing agitated or slowed down

♦ Showing fatigue or loss of energy nearly every day

♦ Feeling worthless, or excessive or inappropriate guilt

♦ Displaying a diminished ability to concentrate, think, or make decisions

♦ Recurrent thoughts of death, suicidal tendencies, or attempting suicide

As with the symptoms of generalized anxiety disorder, the symptoms of depression cause significant distress or impairment in a person's ability to function at work, at home, and in society.

 Under Your Roof _____

If you believe your son or daughter is suffering from anxiety or depression, consult your physician or therapist.

Here are some common reasons why your daughter or son may stagnate and have trouble making progress toward leaving the nest:

♦ As with the scenario just discussed, perhaps your child lacks the skills and tools he or she needs to survive independently. If there's no progress, what's stopping your son? Does he know how to write a resumé? Does your daughter have problems with budgeting or filling out college applications? Talk to your child. Offer your wisdom. Help your child hone the skills needed to leave.

◆ In some situations, a mental health problem like depression or anxiety may paralyze an otherwise healthy young adult. Does your son seem sad, stressed, or stuck in a rut? Does your daughter hole up in her room all day with the drapes closed? Keep in mind that many individuals can suffer from mental health problems even if nothing "bad" has happened to them to "cause" the problem. If you're concerned about your child's mental health, get her some counseling to deal with whatever issue is ailing. With issues as serious as these, it's not possible for your child to simply "snap out of it" or will a change.

◆ Veterans returning from active duty may encounter their own set of mental health issues, which will affect the family's dynamics. If your daughter moves home after serving her country, she'll need to adjust to civilian life again. Returning soldiers may suffer from depression or post-traumatic stress disorder (PTSD), or even show an increased tendency toward violence. People suffering from PTSD are unable to release memories of a traumatic event, and even the slightest cue in their current environment that reminds them of the past event can set off a re-experiencing of the trauma. It's a slow process, but with the love and support of family (along with counseling, if possible), life will return to normal.

The APA notes to watch for these PTSD symptoms, too:

◆ Recurrent and distressing recollections of the event, while awake or asleep

◆ Flashbacks

◆ Insomnia

◆ Irritability or outbursts of anger

◆ Difficulty concentrating

◆ Hypervigilance

◆ Exaggerated startle response

When you and your partner are discussing the proposed length of your child's stay, *negotiate* with her rather than laying down the law. Present a united front with your partner when talking with your adult child, and come up with a realistic agreement (see Appendix A for suggestions) that will work for everybody. Don't be afraid to share your concerns about your child or your feelings about how your child's presence at home affects you (and everyone else in your household!).

Parental Road Blocking

This next issue brings to light a parental misstep. What happens when Mom and Dad don't want to see their child go?

The parents of 20-year-old Beth who moved home for a brief three-month stint between jobs simply weren't ready to see their daughter move out again. Since Beth had been living with her mom and dad again, her parents' marital tension waned. Her mother found renewed purpose in her life. Her father enjoyed helping Beth tune up her old car. They were a "family" again—and her parents weren't ready to give that up quite yet.

To convince her to stay home, Beth's parents threw up roadblocks and incentives every which way, coming up with reasons why she should stay home. Her mom said she needed her to take care of her sore back. Her father said he'd pay her $35 an hour to help him remodel the front room. They tried everything to get Beth to stay.

Beth felt obligated—and guilt-tripped—into staying. And she'd have to admit, staying with her parents gave her perks it would be hard to earn on her own. Her parents' persuasion worked, and Beth remained stuck at her mom and dad's house, working around their house when she should have been working toward independence.

Ground Rules

Helping is when you're teaching and encouraging your child to progress in a way that fosters growth and independence. For instance, you're *helping* your son when you teach him how to balance his checkbook or fill out his financial aid application. *Enabling* is when you provide your son or daughter with resources that hold the child back by preventing her from learning to take responsibility for herself. You're *enabling* your daughter when you support her financially rather than give her the tools to do it herself.

The Analysis

In this scenario, Mom and Dad were blocking Beth's re-launch into the world. They triangulated their daughter, using her as a tension-breaker in their marriage. Instead of working toward her life tasks, Mom chose to take on the "mom" role again. Fearful of being lonely, Dad found a friend in his daughter instead of finding a golfing buddy.

Rather than each member of the family working toward his own age-appropriate life tasks, the parents pulled Beth into their dysfunctional behavior and prevented her from progressing forward.

The Solutions

It's not unusual for parents to consciously or unconsciously hold their son or daughter back. If you find yourself pining for the past or pre-venting your child from moving on, try these techniques:

+ **Examine your motives.** If you're blocking your daughter or son from leaving the nest, take a close look at the reasons why you won't—or can't—let go. Are you using your child as a buffer in a failing marriage? Perhaps you're afraid of being alone or of losing your "job" as parent. In many cases, talking to an impartial friend or seeking marital or family counseling may help you sort out your feelings and come to terms with your emotions.

+ **Take up a hobby—or get a dog.** Parents who dread the idea of being alone should seek out new activities and friendships. This is the time of your life when you should be deepening relationships with your cronies, so search them out. You may even consider adopting a four-legged friend! Studies show that pets do indeed alleviate loneliness, but be sure to find some human pals, too.

+ **Do some soul-searching.** Moms or Dads who feel they have no purpose once their son or daughter moves out should do some serious soul-searching. What do you want the next chapter of your life to look like? Many people at this life stage look outside them-selves, volunteering in their community and giving back in some way. You may even decide to go back to work. Your parenting days are behind you; strive to move into the next phase of your life.

Parental Pity

Sometimes, parents just don't have the heart to kick out their kid, even if she or he really needs to hit the road.

This final case involves Robert, a 30-year-old unemployed computer technician with a young daughter, who moved into Mom and Dad's house—and hadn't left for three years. Robert drank a six-pack of beer every night while his mother cooked his meals, cleaned his room, did his laundry, and cared for his daughter.

Robert's father, frustrated by his son's lack of motivation or skills, had essentially checked out. He often hid in his workshop and immersed himself in his miniature train set, creating his own microcosm. Robert's mom, however, felt consumed by pity and guilt. She felt sorry for her son, and convinced herself that it was her duty to care for her child, no matter how old he was. After all, if he was incapable of being out on his own, wasn't it *her* fault for not raising him right?

The Analysis

Robert's parents knew in their gut that his time to leave had come. They'd supported him through a rough time, they helped care for his daughter, and they even put up with his alcohol abuse. But Robert over-stayed his welcome, and Mom and Dad were ready for him to find a place of his own.

At the same time, Robert's parents—mostly his mother—enabled his lifestyle by cooking for him, cleaning for him, and doing his laundry. She allowed the behavior to continue, even when her husband checked out! She put the needs of her son before the needs of her husband and their relationship. Because she focused her energy on caring for Robert, her bond with her husband deteriorated when it should have been growing stronger.

Guilt played a major role in the cause of this situation, too. Robert's parents may have felt a duty or responsibility to their son, and when he failed, Mom took the blame. She felt guilty about not doing what she was supposed to do.

The Solutions

In this instance, the old Chinese proverb rings especially true: "Give a man a fish and you feed him for a day. Teach a man to fish and you feed him for a lifetime." Parents who want their son or daughter to leave but don't know how may want to follow that sage advice and try implementing some of these techniques:

- **Equip the child to leave.** Offer age-appropriate help to your son or daughter help in a way that's not enabling. Teach the child how to find his or her own apartment. Show her how to do laundry; teach him how to create a budget. By giving your child the skills to survive in the real world, you're giving the hard-earned gift of independence.

- **Help pay for counseling.** If you think your child may benefit from talking to an expert, get therapist recommendations from your family or friends and pay for some counseling sessions. Most therapists don't consider paying for therapy to be "enabling" because it can be too expensive for the child to afford therapy without financial assistance.

- **Toss him out of the nest—in a loving way, of course.** Sometimes, it may take a little push. Kids are often more resilient than parents give them credit for, and they can survive without a mother hen's protective wing. Your young adult's probably more ready that you think!

Parents who feel guilty about kicking an adult child to the curb who is not quite ready should take a good look at if it's *good* for the child to remain home. List the benefits and drawbacks of why he should stay, list the benefits and drawbacks of why she should go. Talk with your partner, and talk to friends and people outside the family for their unbiased opinions. They'll likely tell it like it is!

A Dream or a Nightmare?

For some parents, having an adult child move back home may seem like a dream come true, while for others, it could be their worst nightmare. Whichever is true, you'll need to deal with issues that work to speed up or drag out the time frame for an adult child to stay at home.

When dealing with issues surrounding the length of your son or daughter's stay, most of them can be remedied through family communication, setting up goals and milestones, and negotiating as adults.

The Least You Need to Know

- As you're deciding how long to let your son or daughter remain in your open nest, be sure you, your partner, and your child share the same vision.

- If you don't believe your child is ready to leave the nest, examine why you feel that way. Help your daughter or son to develop the skills needed to survive rather than enabling him.

- Sometimes, adult children who move home suffer from mental health challenges, like anxiety, depression, or post-traumatic stress disorder. Know the signs and get your child help if she needs it.

Chapter 6

Making Room

In This Chapter

- ◆ We all need our space!
- ◆ Revamping the old rules to fit the new open-nesting environment
- ◆ The bedroom: your child's sanctuary
- ◆ Sharing the rest of the house and resources

Whether your child's room has remained a shrine to the best kid (and parents!) in the history of families, or whether the day your child went off to school or work the bedroom became your home office or crafts room, the truth is that you've made adjustments to living in the space of the family home—adjustments that will need to be undone (or at least rethought) when your adult child moves back in.

Making room for your son or daughter involves more than simply inviting your child to move back into an old bedroom, though. You'll need to look at making room throughout the house, from the kitchen to the living room. You'll also need to look at sharing common household items. When your son or daughter moves home, your family (younger siblings included) may even return to its old way of divvying up space, time, and resources.

In this chapter, we look at how to establish territory at home, and how to redefine personal space among family members now that an adult child is living at home again. Space problems and boundary issues will come up, and we'll give you some great strategies for examining the family floor plan and giving each family member the space he or she will need.

Give Me Space

Ecologist and research psychologist John B. Calhoun studied the subject of population density and its effect on human behavior extensively during his career at the National Institute of Mental Health in the 1960s through the 1980s. In 1968, he conducted his most famous experiment, the mouse universe. He brought together a population of mice in one 9-foot-square metal pen with 54-inch sides, with no predators or adversities other than space constraints. The group began as four breeding pairs, and in 315 days, it reached a population of 620. The result was a breakdown of social structure and normal social behavior, and an eventual collapse of the whole population.

Like those mice, we all need our own space. Adults and children alike need a haven to get away from the busyness of life and from the noise and confusion of work, school, family, and friends. We need a place to be quiet, a peaceful refuge to simply get away. Space, in fact, gives us the ability to express ourselves more openly.

Sociologists studied individuals whose ancestors come from northern colder climates and southern warmer climates; they discovered that the cooler-climate people tended to be more conflict-avoidant and less openly emotional, as it's harder to "get away" from family members if there's a conflict. Escaping from conflict would entail venturing out into the inhospitable outdoors. The warmer-climate people tended to be more openly emotional and expressive because the great outdoors could

Ground Rules

Sharing feelings with family members is an important aspect of differentiation, which we discussed in Chapter 3. It also helps young adults become interdependent with other members of the family.

be used as a refuge of personal space. If we extrapolate these findings to an open-nesting situation, when each family member has private space in the house, everyone is more likely to feel comfortable opening up and sharing feelings with others.

Young adults especially need their own space. They're at the point in their lives when they should be coming into their own and becoming independent. A child in the early adulthood phase of life is developing the ability to care for her own financial, emotional, and spiritual needs. When your adult child has a place to be alone, she will accomplish those life tasks—and fly the nest ready to face the world.

Think about it. When you were 23 years old, you likely enjoyed retreating to your room, cozying up on a comfy chair and writing in your journal or enjoying the privacy to chatter on the phone to friends about your latest love interest, particularly if you had roommates or lived at home. It was your place to relax and be alone.

While not every household can provide a retreat for a returning son or daughter, it's helpful for you to realize the significance of that space to your child. Young adults are in the process of emotionally separating themselves from their parents, and they need to be given the physical room to do so.

Recalling Calhoun's "mouse universe," the house (or pen) itself makes a big difference when it comes to giving your adult child the space he needs. The larger the better, in most cases, but families with limited room can make open nesting work, too. It just takes some creative coordinating and arranging to make it happen.

Then Versus Now

Sharing space with your child is nothing new. When your daughter or son was growing up, the family lived together under one roof. How did your family share the space? Perhaps your son rarely ventured into the kitchen because Mom or Dad prepared the meals. Maybe the big sister bullied the younger sibling into taking the least-desirable hours with the Playstation. You may have tolerated your son's dirty socks, soccer cleats, or football pads (but there's no way that will fly now!).

Now that your child is moving home as an adult, you will need to update those old rules and structure. You can't force your son or daughter into the juvenile concepts of "you're living under my roof and will do what I say." As much as you may want to, you can't dictate strict rules about how clean your daughter's room should be, or that your son must take out the trash every night. Those rules will need to be let go for your adult child.

But you can learn how to co-exist as adults, respecting each other's physical and emotional space. But it will take some getting used to, that's for certain. Let's take a closer look at each section of the house and explore some possible issues that may arise.

The Bedroom

When Jane, a 23-year-old financial consultant, moved back into her mom and dad's home after graduating from college, she had it made. Inside her parents' spacious five-bedroom house, Jane settled into an expansive bedroom with its own sitting area, bathroom, and staircase leading outside. She could come and go and entertain as she pleased.

Kevin, however, didn't have it so lucky. When the 28-year-old soccer coach returned to the nest after six years of living on his own, his parents had remodeled the house, knocking down the wall in his old bedroom to expand the kitchen and living room. He had to bunk with his kid brother for the first few months until he took over his younger sister's room when she headed off to college.

As with these examples, your living situation and the space you have for your child will be unique to your family. Perhaps you kept your son's room exactly as it was when he left. Maybe your younger daughter moved into your adult son's room when he moved out, and now that he's returned he's left with the pink frilly room abandoned by his sister. You may have turned your daughter's room into a den or craft room, or even downsized from a four-bedroom house to a smaller two-bedroom condo.

When you're trying to figure out how to give your adult child a room of her own, consider these points:

◆ **Decide what you're willing to sacrifice.** Look at your situation and decide what lengths you'll go to allow your child to move in. Are you willing to temporarily give up your sewing area or music room for your daughter? Do you need your office to work from home? You may need to draw the line and offer your son the smaller guest bedroom (or garage) instead. You don't need to inconvenience yourself, but if you do make some sacrifices, you should remember that it should be a short-term arrangement.

♦ **Consider what you're willing to invest.** There could be some cost involved with moving your son or daughter back in. You may need to hire movers to clear out the room or Salvation Army to haul away the Barbie house, Tonka trucks, or the old school projects. You may have to rent a storage unit to store your furniture, whether it be the treadmill or the antique writing table. You may need to paint those pink walls a softer shade of beige. How much are you willing to do … and spend?

♦ **Check your feelings.** Don't make any decisions that cause you to feel like you're being taken advantage of. If you're not ready to give up your painting studio, don't! If you feel like you're sacrificing too much, look at ways to compromise. Remember, you're doing this out of love as a favor to your child, and you don't want resentment or negative feelings to taint the sentiment.

Ground Rules

When it comes to chores, pick your battles. Family therapists often hear parents complain about their child's messy room. Lauren advises her clients not to nag their teen or adult child. Rather, parents should think of the child's room as "her space" to keep as chaotic or tranquil as she wishes, and encourage her to do her part to keep the common areas—the ones that everyone uses—neat and tidy.

The Kitchen

Lindsey, a 26-year-old cosmetologist, loved to cook but, wow, did she make a mess. When she moved home, she spent nearly all her free time in the kitchen, whipping up fancy sauces, baking buttery breads, and making decadent desserts. In her wake, dirty dishes overflowed in the sink, flour dusted the countertops, and who-knows-what lined the microwave walls.

Mom and Dad didn't want to squelch Lindsey's culinary enthusiasm, but the messes soon became too much to bear. Lindsey's parents invited her sit down, praised her cooking skills, and thoughtfully explained that they preferred to have a clean kitchen. Lindsey understood. From that point forward, she continued with her hobby—but added cleaning to her routine, too.

The kitchen is the heart of the home. It's where the family members come together to cook and eat. It's where food is prepared, stories are shared, and midnight snacks are munched. In an open-nesting situation, you'll need to make room for your son or daughter in the kitchen, too. The child will need to cook and eat, after all, so she will need space in the kitchen to do that, even if you're cooking the meals.

It's very likely that you'll be sharing appliances, like the refrigerator, stove, and microwave. Your child may use your plates, cookware, utensils, and Tupperware. You'll also be sharing space on the shelves and in the fridge.

When you share space and resources in the kitchen with your adult child, you may run into these culinary complexities:

- **Running into each other.** A kitchen is only so big, and if your son or daughter is making meals, mealtime can mean chaos in the kitchen. Plus, if your daughter invites friends over for champagne Sunday or if your son hosts a playoff party, you may wind up with too many cooks in the kitchen. If this becomes an issue in your household, consider scheduling kitchen time—or at least sharing plans with each other.

- **Missing food.** You've been looking forward to that last slice of chocolate cake all day, but when you open the refrigerator door, it's gone! When your son or daughter moves back home, your goodies may vanish if they're not clearly labeled. If disappearing food becomes a problem in your household, talk to your child about it. Put your name on what belongs to you, if necessary. If your child tends to gulp the last swig of milk, establish the rule that if someone uses something up, that person is responsible for replacing it.

- **Your shelf, my shelf.** If you and your child decide to purchase your food separately, you may want to consider giving her a designated shelf in the refrigerator and a section in the pantry. That way, there's no cause for confusion—and accidental ingestion.

◆ **Clean-up duty for all.** You'll be sharing the kitchen space and the tools—but hopefully not the mess. Your son or daughter may need a gentle reminder to put dirty dishes in the dishwasher when he's done with them. Be sure to discuss your expectations for cleanliness and tidiness in the kitchen, and if conflicts arise, use a relational approach rather than punishing your child.

The Living Space

When Carl, a 24-year-old landscaper, moved in with Mom and Dad, the house went from neat and tidy to downright dirty. It seemed like Carl brought a wheelbarrow full of dirt home from the job sites every night. He tracked mud from the garage to the bathroom, leaving a trail of filthy t-shirts, handkerchiefs, shoes, and socks behind him. It didn't take long for Mom to lay out her standards for cleanliness in the common areas.

Common areas, like the living room, the bathroom (unless your adult child is blessed with her own bathroom), the laundry room, the back yard, and even the garage may need to be shared, too. Your child is moving into your home—not just her bedroom—and you and your partner will want to make sure your adult child (and any or all siblings) understands what's expected.

You've shared space with your son or daughter before, but now that he's home again, you'll need to treat your child like the adult he is. Here are some great suggestions for handling situations that may arise:

◆ **Communicate often.** First and foremost, talk with your child about sharing common space. What are your standards for cleanliness and tidiness? You may be okay with a little dust but not strewn socks and stinky sweatshirts. Do you have routine that you'd like to keep? Tell your child that you do your laundry every Tuesday and you expect the washer and dryer to be cleared out for you. Sharing your thoughts and feelings about the situation will let your child know how to integrate back into the household. And make sure to tell your adult child as soon as house-sharing habits bother you—like his letting the teapot blast its boil or her constant tone-deaf singing. Speaking up will prevent you from storing up resentment.

◆ **Request a heads-up.** It's okay to ask your son or daughter to forewarn you when bringing guests over. When your child lived on his own, for example, friends may have come and gone at will, but now that he's living in your house, he should consider it common courtesy to give you a heads-up before inviting over a pack of 20-somethings to drink beer and play a six-hour video game marathon.

◆ **Create schedules.** When all else fails, you may consider creating a reservation calendar for using various common areas around the house. It doesn't have to be complicated. Simply print out a blank calendar and ask family members to jot down the room and hours they'll need it. For example, every Tuesday, Mom would write, "Laundry, 9 A.M. to 2 P.M." Or, if you and your wife want a stay-at-home date night, you'd write, "Family room, 7 P.M. to 10 P.M."

Under Your Roof

Worried about imposing? Don't hesitate to ask your son or daughter to pitch in if need be. Rather than pay for movers or painters, ask your child (and his strong friends) to do the heavy lifting or painting. If the child is able, he could also pitch in financially. It'll help make the room his own—for the time being.

Sharing Technology

Technology and computers have taken over our households, making many of our lives easier and more efficient—and a lot more fun. We have access to the world through our keyboards. We can hold our entire music collection in the palm of our hand. We can play chess against someone across the world. We can even enter a virtual world and shred moguls on a snowboard, practice a golf swing, or hit a baseball out of the park without even leaving the house.

You likely have all kinds of electronics in your house, from a family computer and video-game console to the flat screen, DVR, and stereo system. When your adult child moves back home, you may need to share these luxuries as a family. Many fortunate young people today have their own laptops, game consoles, televisions, and MP3 players, but you might find yourself bickering over a television show once in a while.

To ensure all the people in the household (including siblings) have access to the electronics they treasure, try these tips to make it easier:

◆ **Give everyone his or her own multi-media hub.** If your family has the financial and logistical resources, consider outfitting each bedroom with its own television, DVR, DVD player, computer, and Internet access. That way, if your son wants to update his Facebook status while his sister is typing up her research paper, he can head to his room instead of harassing her. If that's just not possible, consider creating a schedule. Share when you can, and when you can't, develop strategies for making sure each person gets needed time to do important things.

◆ **Compromise.** Sometimes family members will simply need to give and take. You may have to record your favorite television program once in a while. Your daughter may have to check her e-mail later after her sister is done using the computer. It's a matter of communicating needs or wants, sharing resources, and giving in sometimes.

Ground Rules

Your adult child should be mature enough to negotiate and compromise with your younger child for time with electronics. Even if you hear them arguing, you should only intervene as a last resort.

◆ **Consider the other family members.** In an open-nesting family with younger siblings, parents should be very careful to respect the rights of the younger sister or brother. Before your adult child moved home, the younger sibling had all the access she wanted to the family computer and television. Now, kid brother or sister has to share, and that could cause conflict in the household if you don't ask your adult child to mitigate it.

◆ **Create a formal schedule.** Consider creating a formal schedule of some kind. Just as we suggested with reserving common areas on a calendar, do so with the various electronic equipment. Johnny, for instance, may book the computer from 1 P.M. to 4 P.M. on Thursday. Mom may reserve the Wii every morning to practice yoga. Lisa and her friends reserve the television every Monday at 9 P.M. to watch *24*. A written schedule can facilitate communication between family members, especially in a busy household. Make sure that all family members are present as the schedule is being created. Use a relational approach (see Chapter 3) to ensure that everyone's needs are being heard and taken into account.

By making sure each family member has some territory to call his or her own, everyone will feel freer to share their thoughts, feelings, and emotions—and feel more comfortable in this new living arrangement.

The Least You Need to Know

◆ We all need our space. As individuals, space helps us get away from the busyness of life. In a family unit, it promotes bonding and sharing of emotion.

◆ The rules and structures your family used to define space when your child was young will need to be revamped now.

◆ When deciding on a room for your adult child, decide what you're willing to sacrifice. If your son's old room is still open, great! If you made it your den, are you willing to give it up?

◆ Beyond your child's bedroom, be prepared to make room in the kitchen, the living room, the bathroom, and even the garage. You also may need to share electronics, too.

Chapter 7

Three's Company ...
Is Four a Crowd?

In This Chapter

- ◆ Handling conflicts and learning to communicate with your child's partner

- ◆ Blending two families

- ◆ Common challenges when your child moves back and brings a partner

- ◆ Letting your child go—even though she's under your roof.

It's one thing if your adult son or daughter moves back home. But what if your child's partner (and maybe even their children) settle into your nest, too? Sometimes open nesting means two families need to come together under one roof—and it's happening more and more today. Whether it's a military family moving off base, a newly married couple struggling with bills, or a young family saving for their first home, many extended families are seeing the benefits of coming together, at least for a short while.

The situation presents some challenges for both you and the young family. You may not have much space for two or more extra people in your home. Your child's partner may not get along or feel comfortable living together as an extended family. Your traditions and habits may differ from those your son or daughter's partner is accustomed to. The open-nesting environment may hinder the new couple's developmental task of forming a primary bond with one another.

It can be successful as long as the parents and the young family understand that their relationship in their primary family—as devoted partners—takes top priority.

Conflict Resolution

When it comes to keeping the peace in a house, more often than not, women are up to the task. Traditionally, moms, and sometimes Grandmas, are viewed as the emotional center of the family, smoothing over conflict, nurturing sons and daughters, and helping everyone simply get along. They are often the ones who weave together everyone's lives into a "nest"—and keep that nest strong and sound. How many times have you (or your wife) separated squabbling brothers or eased your daughter's hurt feelings?

A Failure to Communicate ...

Jose and Sara, a young married Latino couple in their mid-20s, moved in with Jose's family to save money for a new condo. At first, things went well. Everyone got along and respected each other's space, but when Jose's mother started to hint to her daughter-in-law about finding a job, Sara started fuming.

Rather than talk to her mother-in-law and ask that she refrain from leaving help-wanted ads on the kitchen table and e-mailing job listings to her, Sara vented to Jose. It put him in an uncomfortable position: Jose wanted to be a good husband and ask his mom to cool it, but he also wanted to be a good son and show respect to his mother—especially because she was opening the nest to put them up.

In the end, the family resolved the conflict during a family meeting, which they decided to hold on a weekly basis. Sara checked her emotions and gently expressed her frustration with her mother-in-law, who was unaware that her "helpful suggestions" were perceived as nagging. The two women worked through the hiccup and grew closer because of it. Jose was relieved to stop being the emotional go-between for the two most important women in his life.

And Junior's Partner Makes Four

When a partner or spouse is added to the mix, Mom (and Dad, too) may be affected if the son- or daughter-in-law hesitates bringing up issues or conflicts. For instance, if a new daughter-in-law feels uncomfortable approaching her in-laws about them invading her privacy, she may tiptoe around the issue, be indirect about her concerns, and be overly stand-offish. The daughter-in-law may then develop festering feelings or put her husband—your son—in the middle as a go-between, sparking undue anxiety.

Similarly, the parents-in-law may have qualms about asking the son- or daughter-in-law to change how he or she does things. Rather than talk to the son-in-law directly about not cleaning his dirty dishes or the daughter-in-law about leaving her hand-washed bras hanging in the bathroom, the parents may pull in their son or daughter to act as conflict-solver. This triangulation will put your adult child in an uncomfortable and stressful spot.

Under Your Roof

Triangulation, which is when an individual is used as a tension breaker or go-between in a relationship, can occur between more than just three individuals. It can also happen in group interactions, like when parents pull an adult child between them and her partner.

If you find yourself unable to mitigate or resolve conflicts with your in-law child, try to incorporate these strategies in your day-to-day dealings:

♦ **Hold family meetings.** Once a week or every two weeks, have a family meeting with your child, her partner, and any younger siblings living at home (if they're old enough to participate in a constructive way). It can be a formal sit-down, or it can be a casual conversation over dinner. Discuss how the living arrangement is working and encourage each family member bring up any requests or issues that may have surfaced. So that everyone will be emotionally prepared, however, set the expectation that some issues will need to be discussed and resolved. Keep strong emotions off the table and discuss the matter like adults. (For more information about family meetings, please flip to Chapter 17, where we provide a detailed "how to" on conducting constructive get-togethers.)

♦ **Communicate proactively.** You can make a special effort to foster conversations and dialogue with your child's partner. Get to know her. Learn about each other. Ask what preferences your son-in-law may have. Communicate directly with your daughter-in-law. Be curious and be open. If you have a comfortable back-and-forth, maybe any thorny issues will come up (and be resolved) on their own.

♦ **Don't put the adult child in the middle.** If need be, your adult child has the right to refuse to be put in middle by asking family members to communicate directly with each other. If conflict does rear its head between parents and their children's partners, remember: It's critical for the adult child and his partner to create an executive system—in this case, two partners who make decisions together and support each other first and foremost. And if partners don't agree—either yours or your child's—then take the role as your partner's advocate and do your best to mediate a good solution that makes everyone happy (or at least, happier!)

Under Your Roof

Family meetings are intended to foster communication, but they're also a time to bond and share time together; it's really the process that's important. Make the most of it by staying positive and working toward solutions. Encourage thorny issues to be presented as opportunities for problem-solving and a chance for your family to grow. This helps keep meetings from becoming rants and keeps everyone focused on finding answers, not on placing blame or provoking screaming matches.

When conflicts come up—and it's likely they will—keep the lines of communication open between all family members. Disagreements are normal, but when bad feelings start to fester, resentment may set in. Simply talking to your son- or daughter-in-law directly may ease some of that tension. (Look at Chapter 11 for some communication strategies to try.)

Blending Well—or Not

Many times, adult children choose spouses or partners who're compatible with their own family of origin. Your daughter or son may naturally gravitate toward a person who shares the same values, religion, personality type, and outlook on life. When the couple moves home to either partner's folks' home, the family has a head start toward getting along.

But sometimes, the new partner doesn't get on so well with the rest of the family because the adult child chose someone completely different from his or her own family of origin. The son may choose a woman from a different religious background. The blueblood daughter may choose a blue-collar man with no college degree. Shakespeare's *Romeo and Juliet* exemplifies this point precisely: the way Juliet chose to distance herself from her family (and develop her independence) was to choose Romeo, someone forbidden who looks, acts, and believes in something completely different. Parents can see this as a direct challenge.

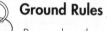 **Ground Rules**

Remember the phrase, "Pick your battles." Broach only the biggest issues rather than nitpicking. Your child and her partner's stay is only temporary, and there's really no need to harm your relationship with them over sweating the small stuff.

If this describes your open-nest household, try to stay open to your in-law's differences and accept this young man or woman your child has chosen. Though it may be easy to see this person as a threat (you're just protecting your child, after all!), try to view him as someone your child loves and plans a shared life with. If you can accept him, chances are good that he will also accept you.

Try these tips to develop a closer bond with your adult child's partner:

♦ **Find common interests.** Your adult child is only one of many common interests you can find if you try. When the moment presents itself, ask your in-law (nonjudgmental) questions about his or her life experiences. Ask about hobbies and interests. Be curious and open-minded. Who knows? You may learn that you're both avid baseball-card collectors or Sudoku buffs. You won't learn unless you ask.

♦ **Find out why your child fell in love.** When you see your son or daughter's partner through your child's eyes, you may come to appreciate this individual—and even learn something new about your child! Ask your child what made this man or woman *the one*. Doing so will generate positive feelings about the person, and you may start to see beyond your differences.

♦ **Celebrate your diversity.** Sure, you're different, but why not see those differences as an opportunity? Learning about your son- or daughter-in-law's unique perspective on life may give you a chance to expand your own horizons. For instance, if your son-in-law is a NASCAR fan, try to learn a thing or two about Jeff Gordon. Likewise, if you're a bluegrass fan, maybe your son- or daughter-in-law can learn to appreciate Laurie Lewis's fiddle-playing ability. Try to stay open to exposing yourself to new—and different—things.

♦ **Keep it positive.** If you and your in-law don't get along at all, resist the temptation to speak out negatively—especially to your adult child. You may want to highlight the "bad" things, but that will only make your child feel uncomfortable. As the old saying goes, "If you can't say something nice about a person, don't say anything at all."

Sometimes people get along, and sometimes they don't. In an open-nesting environment, it sure makes things a lot easier when everyone is amicable, but even when they aren't, living together can work, too. Remember to keep the interactions as positive as possible and try not to put your son or daughter in the middle of conflicts. Keep the focus on your extended family's goal in coming back together; accomplishing that goal will only make the whole family stronger and happier.

"Nesting" in the Nest

To help your adult child and partner feel more comfortable in their temporary digs, try these techniques:

◆ **Remember that paint is temporary.** Invite them to paint and decorate an area of the house, like their bedroom, as they wish. If they have their own furniture, encourage them to move it in and arrange it however they want. They may even want to change the window treatments. Let them do an "extreme makeover" in their room.

◆ **Mentally move ahead.** While they're living with you, encourage them to look toward their next life stage—moving into their own place—and help them formulate goals to get there. They may want to go online and explore where they'd like to live and what to put in the house. They can even start purchasing *their* own furniture (as long as you have room in the garage, of course!). Help them plan and set goals that keep them excited and working on their future together.

◆ **Rewrite the script.** If embarrassment is getting your young marrieds down, remind them how many young people live at home with their parents at some time in their adulthood. The "script"— get married, buy a house, have kids—is changing. Chances are that living at home will give your kids a leg up they'll need to launch into a living situation they can be proud of and sustain without too much struggle.

Any young couple will feel the need to nest. You can allow (and even pay for) these young lovers to create their own space together, which will help everyone to feel more comfortable in a possibly uncomfortable situation.

A New Generation

Newlyweds Paul and Kathy—and their Bichon Frise named Daisy—moved in with Paul's mom and stepdad, Bob and Elaine, to save up for a condo of their own. Paul's parents weren't pet people and their home wasn't dog-proofed, so the introduction of this new four-legged family member proved challenging indeed (especially after Daisy's first accident on the Persian rug!). After some compromises, like sequestering Daisy in the kitchen and in Paul and Kathy's room, Paul's parents accepted the little canine fluffball, even allowing some supervised play time in the living room. And wouldn't you know it—after Paul and Kathy moved out, Bob and Elaine adopted their own pair of Bichons!

Get a Room!

Besides the potential conflicts and not being able to call a house truly *home*, a young adult couple may also struggle with finding time to be private—and intimate—with each other while Mom and Dad putter around. The flip side of the coin may be true, too. You may feel uncomfortable being intimate with your partner while your adult child or her partner is home.

To keep any married relationship healthy, a couple needs time and space to be intimate. It gives them the opportunity to bond emotionally and physically in a way that fosters their unity. If your young marrieds—not to mention you and your spouse—don't have that intimate time together, the relationship may suffer. To prevent that from happening, consider these suggestions:

◆ **Schedule at-home date nights.** To give you and your young couple the space and privacy everyone needs, arrange for each couple to be home alone a certain night of the week. Tuesday may be your night when you reserve the "house" from 6 P.M. to 9 P.M., during which time your child and his or her partner can enjoy a movie or a night on the town. Friday may be the young couple's night, giving you and your partner an excuse to try that new sushi restaurant down the street.

◆ **Set up house rules.** You may also want to create some rules with regard to privacy, like "no knocking on a closed door," or "no entering the other couple's bedroom." That way, everyone will feel comfortable and no one will worry about accidentally seeing more than necessary. Locks on bedroom doors might be a good idea, too.

Intimacy is important in any relationship, and in an open-nesting household, you may find it challenging to get that alone time with your partner. Use your family meetings to come up with some creative solutions that enable both couples to have their private time. It'll keep everyone's spirits up!

> **Under Your Roof**
>
> If you're uncomfortable with your child being intimate with a partner under your roof, it's okay for you to state your feelings about it. However, try to make peace with the fact that the couple has reached adulthood, and the pair will ultimately do what they wish.

Parents Who Won't Let Go

If your family tends to share a little too much with each other, keep these points in mind, especially if your child moves home and brings a partner:

- **Create couple respect.** Each couple in the household should form their own executive system with a firm boundary around it that keeps outside influences and opinions from infiltrating. Think of these systems like mini-countries: you and your partner form one country and your child and his partner form another. When decisions are made, you do so within your own boundaries. The United States doesn't consult with Chile when deciding whether to lower taxes!

- **Consider your own partner first.** Remember that the couple is now the primary unit, not the nuclear family. Both couples need to turn to each other to make decisions. No matter how tempted you may feel about weighing in on whether Junior should take that job, let the young couple decide on their own. Sorry, Mom and Dad: you're not the boss any more!

- **Hold on to yourself.** Try to manage your own anxiety when you worry that your son or daughter is making the "wrong" decision. It's hard to sit back and watch someone you care about do something you disagree with, but it's inevitable that this will happen with your adult child. Keeping your mouth shut will cause you anxiety, so you'll need to manage it with some anxiety-reduction techniques. Try listening to relaxation tapes, doing some deep belly breathing, meditating, taking a long walk, or enjoying a hot bubble bath. Talking it over with your own partner can also be good … but not if it creates even more anxiety! Look for other outlets.

It's okay for family members to be close to one another. But when that closeness comes between intimate partners in a marriage, it prevents them from moving through this developmental stage in their relationship. It's time to let go, even when this precious young couple is living back in *your* nest, right under your nose!

The Least You Need to Know

- When your adult child's spouse moves back to your home, too, you have a unique opportunity to get to know your in-law very well—for better or worse!

- Regular family meetings can mitigate problems before they occur and foster an environment of healthy communication.

- Young couples will want to "nest" in your open nest—so let them! Encourage them to bring their own flair to the family décor, even if it's just in their own room.

- Though you're living under one roof, it's critical for newly marrieds or domestic partners to continue working toward their developmental tasks as a young couple.

Chapter 8

Dating and Sleepovers

In This Chapter

◆ Recognizing that sex is normal for your adult child, even though it's a taboo topic!

◆ Identifying common issues surrounding dating and sex in the open-nesting environment

◆ Exploring issues for dating parents, too

◆ Feeling comfortable communicating is key

So you thought parenting an adolescent child's passage through puberty was the most challenging thing you've done? Think again! When your adult child moves back in with you, you'll need to explore issues of sexuality and dating all over again.

As parents, you'll have a different role to play in your adult child's relationships. You won't have a say in who your son or daughter dates. You can't forbid your daughter from meeting guys at bars, or chastise your son for not calling that "nice girl" back. You'll need to remain more or less as observers while your child searches for the perfect partner (or has fun dating lots of different people).

If you're a single parent, you may be dating, too. Your adult child will be privy to your comings and goings and may have an opinion (or two) about the partners you're choosing. Your child may compare your new partner with your ex, or may not be emotionally ready for you to date another person. But just as you should hold judgment, your son or daughter should, too.

When it comes to sex, intimacy, and dating, the most important thing is to be open and honest with adult children. Address any issues that surface with candor and honesty. Rather than sweep the issue under the rug, make certain everyone's feelings are clearly, appropriately, and realistically communicated.

Are You OK With It?

According to studies by the Centers for Disease Control and Prevention, in 2002, the majority of young adults—70 percent of women and 65 percent of men—had engaged in sexual intercourse by age 19. Couple that with the fact that young people are marrying later in life—25 for women and 27 for men, according to the U.S. Census 2004 figures—and it's pretty clear that many young people are having sex out of wedlock.

Some parents, however, may not feel comfortable with the idea of sex before marriage. Maybe they waited until their wedding night to have sex. Perhaps they have a hard time imagining their child having sexual relationships with more than one partner. In other cases, there may even be a double standard, where the parents engaged in premarital sex but expect differently from their children.

A New Generation

Surprisingly (and encouragingly), statistics from the Centers for Disease Control and Prevention suggest that in recent years, more young people are delaying their first sexual experience. In 2002, 13 percent of female teens had sex before age 15, compared to 19 percent in 1995; 15 percent of males had done so, compared to 21 percent in 1995. The reason seems to point to changing morals. In the same study, among teens who had not yet had sex, the main reason for being abstinent was that it was "against religion or morals," followed by "don't want to get (a female) pregnant."

When your adult child moves back home you may want to take inventory of your own feelings and attitudes about dating, sex, and sleepovers. Are you okay with your son having three girlfriends at the same time? Will you be comfortable with your daughter staying overnight at her boyfriend's house? You likely raised your child with certain morals and principles, but they may not match current cultural expectations, which you'll need to come to terms with.

A Taboo Topic?

In many U.S. families, sex is a "taboo" topic, or simply an embarrassing subject kept hush-hush. Unlike other cultural groups, like the Western Europeans who more freely discuss intimacy and sexual relations, Americans—with our Puritanical roots—tend to talk *around* the subject. There may be a lot of sex on television, but as a topic of conversation among family members, sex is not something that's out in the open. It's something that stays in the bedroom—and even there it's not discussed too much.

Families deal with the topic of sex in different ways. Some, who take the "don't ask, don't tell" approach may assume that sexual behavior is or is not taking place, but they're too embarrassed or maybe ashamed or fearful to talk about it openly. There's an unwillingness to think about or accept the fact that other family members might be sexual people. There may even be a tendency to *ignore* the fact that anyone in the family may be sexually active, or may even *pretend* that it's not going on. It's like when a child thinks of her mom and dad having sex, a common response is "Ewwww! My parents don't do *that*!" Or when Mom finds smuggled birth-control pills in her daughter's dresser, she pretends she didn't see them.

> **Ground Rules**
>
> Intimacy is a valuable and desirable relationship quality to encourage in your adult child's life. Keep in mind that intimacy involves more than sexual relations. It encompasses closeness and care-giving, open communication of feelings and needs, and a deep respect for the other person.

In other families, like the families with a more rigid style we discussed in Chapter 4, parents may try to exert more influence over their child's sex life and state very strong opinions (overtly or covertly) about their values. They may make an instant judgment about their child's behavior without hearing the facts, or they may even try to regulate or control their child's behavior. For instance, parents of a libertine young man may lecture him about his amorous activities, or parents of a young woman who can't ever seem to be home before 2 A.M. may restrict her car privileges in an attempt to quell her late-night exploits.

Does this sound familiar? Are you embarrassed to talk about sex with your family members or even broach the topic with friends? Or do you express your opinion—even when it's not sought? You're not alone. This *is* your "little boy" or "little girl" we're talking about here! Just remember that sex and intimacy are *normal behaviors* for men and women. Biologically, sex is essential to reproduction. Emotionally, intimacy is essential to feeling loved and wanted. They're both desirable—so don't be afraid to acknowledge it or maybe even talk about it!

Accept It!

Like it or not, your adult child will very likely be sexually active and intimate with one (or more) partners. Here, we list the most common concerns that face parents in an open-nesting environment. One theme you'll see throughout the following examples is *openness*. No matter the concern, it's essential that you address it honestly and directly with your adult child.

Sleepover Stress

Leila, a 23-year-old college student, moved in with her mother, Helen, after she and her roommates could no longer afford their rent. Leila had a bustling social life. She frequented nightclubs and bars after working her after-school job as a barista, often staying out all night with her friends.

Helen couldn't stand it. She constantly worried, wondering where Leila was, what she was doing, and whether she was safe. She couldn't fall asleep when Leila was out and fretted that her only daughter was up to something nefarious or in some kind of trouble. Being a single parent, Leila's mom felt an overwhelming urge to protect her child.

If you're like Helen, you may worry when your adult child spends nights away. Those feelings are normal, regardless of your child's age. Parents often have an innate urge to protect their offspring—and that means being concerned about them.

If you're concerned about where your adult child is and what she is doing, ask! Though you would be intruding if you called your daughter when she is out and about, choose a time when she is home, sit down with her, and share your feelings. Helen could let her daughter know she understands that her social life is personal but would rest easier if she knew generally where Leila tended to go, and whom she was with. If you act open and nonjudgmental, your child will be more likely to give you a straight answer. Bring up any concerns you have about safety or well-being, but acknowledge that your child is an adult.

 Ground Rules

Resist the urge to try to prevent your child from spending the night at his or her partner's house, as it is developmentally normal to do this. Plus, your son or daughter is too old for you to be making rules! Your child will probably find a way to get around the rules, anyway.

If you've already talked to your son or daughter about those late-night excursions and still lie awake worrying, try these techniques to ease your mind:

♦ Write down a list of worries, and next to each worry write down several "helpful thoughts" that counteract the worry. Study the helpful thoughts, and acknowledge their truth. For instance …

Worry: I worry that Michelle is going to get pregnant out of wedlock.

Helpful thoughts: Michelle has been taking birth-control pills for the past three years.

She is an experienced and responsible young woman, and she knows the risks and consequences of having sex without protection.

If the worst happened and she were to get pregnant, we have the means to help her raise her child.

♦ Buy a relaxation tape and listen to it before bed. Practice meditation to clear your mind, or say a prayer asking for peace and protection.

- Focus on your own life and activities in order to take some focus off your adult child. Perhaps you could benefit from some new hobbies or more time spent with friends.

- Remember that you didn't have these worries when your adult child lived independently (and your child still survived!).

If these strategies are still not enough to ease your mind, consider setting up a few sessions with a professional counselor (preferably a family therapist), who can talk to you about your specific concerns.

By practicing some anxiety-relieving techniques and communicating with your adult child, your concerns will slowly lift. Before your daughter moved back home, she took care of herself just fine. When he lived on his own, you trusted your adult son to be the responsible person you raised!

Bringing Partners Home

Brian, 26, and his girlfriend, Nina, had been dating for six months when he returned home to live with his parents. Brian and Nina were sexually active—and continued to be when he moved back into his old bedroom. They didn't see the harm in having some intimate time together, as long as the bedroom door stayed closed.

Brian's parents didn't approve of their son's premarital sex, and they most certainly didn't want to know about it if it *was* happening. Brian and Nina's behavior made his parents feel extremely uncomfortable, especially when they saw Nina's car in the driveway at 6 A.M. for the first time.

Because young adults will engage in intimate and sexual behavior, you can expect your son or daughter to bring partners home. Here are a few ways to make it easier on yourself:

- **Accept reality.** Your son or daughter has a sex life. When your child was in high school or even college, you may have cracked down and forbidden that behavior, but now that she is an adult, things are different. Sex is happening, and this time it's right in your house! Chances are that you won't like it if your child brings her partner home—and she won't be too crazy about it, either—but it very well could happen. Try to accept this as developmentally normal.

♦ **Set ground rules.** It's best not to forbid your adult child from bringing partners home (although it's within your rights to do so), but it's okay for you to ask for certain conditions. Is it disruptive when your daughter slams the door at 1 A.M. when she and her boyfriend arrive? Tell her! Would you rather not have your son's girlfriend sitting at the kitchen table at 8 A.M. as you come in to make coffee wearing a bathrobe? Request it! Do they take over the living room all day Saturday to watch movies? Let them know (preferably ahead of time) when you'd like to use the room. Open communication here is the key—even if it might be an "embarrassing" subject.

If you're uncomfortable with your child bringing partners home, talk about it using a relational approach, which involves negotiation rather than hard-and-fast rules. It's within your rights to share your concerns and request a compromise, like requesting that your child and his partner go elsewhere. Your child may even oblige—you're more or less trying to accept what your son or daughter is doing, after all!

An Unacceptable Partner

From the moment Kim's father met her boyfriend, Chad, he didn't like him. Kim and Chad met at a virtual bowling alley on Second Life and they quickly fell for each other after meeting in "real" life. Kim, 24, was attracted to Chad's comedic, frat-boy behavior, even though Chad was 28. Chad's antics completely insulted Kim's father, who raised his daughter in a strict household. She knew her parents wouldn't approve of her new love interest, but she felt a certain freedom in having someone around her who proved unacceptable to her parents.

> **Under Your Roof**
>
> Resist the urge to judge a suitor your child chooses. Often, even a "perfect" partner for your child will differ in some important way from her parents. The choice may even be an attempt by the young adult to display her independence from her parents.

Kim's father voiced his opinion—loud and clear—to his eldest daughter. He wanted her to meet a grounded and caring gentleman, not a rough-and-tumble type with little or no ambition and no apparent interests other than dating his daughter and playing games on the Internet.

Despite feeling tempted to forbid their romance, Kim's father knew that he had to let her figure it out on her own.

As Kim's father learned, parents often have little direct influence over their child's choice of partners. Your child is an adult, and as such, has the freedom to date whomever strikes her fancy. Sometimes, that person will be completely opposite from whom you would deem acceptable.

It's best to be as amicable as possible, because it's likely that your adult child will eventually "drop the zero," anyway. Your daughter may be expressing her individuality and independence by dating an outsider. Your son may wish to hang out with someone with a different background than his own. Putting yourself in the middle sets you up for extra conflict as your child defends his choice, while the new boyfriend or girlfriend just knows you're not happy.

Revolving Door of Partners

Zack was a lady's man. Almost every night, the 27-year-old laid-off customer service representative hit the club scene and played the field, sometimes romancing two or three women at a time. When Zack moved in with his parents, it didn't cramp his style one bit. He still wooed the ladies—much to his parents' chagrin.

After several weeks of watching their son's revolving door of partners, Zack's parents had enough. They were concerned about their son's inability to handle long-term intimate relationships. Plus, they were feeling like their privacy—and safety—was at risk. They didn't know who these ladies were or what they were up to.

Zack's parents had reason to be concerned. Their son's promiscuity went hand in hand with low self-esteem. He was seeking approval and acceptance, and the only way he felt he could do that was through sexual exploits.

A person who suffers from low self-esteem views himself as inadequate, and often becomes an expert at hiding his feelings. Negative thinking tends to control feelings and behaviors. Here are some telltale signs of low self-esteem:

- ◆ Makes negative "I am" statements and apologizes excessively

- ◆ Is eager to please others, and has a strong need for approval and constant support

- Is overly sensitive to criticism but overly critical of others, often putting others down

- Has a strong need for material possessions

- Shows reluctance to express his own ideas and lacks belief in himself

- Feels hopeless

- Fears new experiences and changes

- Strives to be perfect but sees herself as far from perfect

- Exaggerates successes to mask feelings of inadequacy

- Is excessively anxious

If you feel your child suffers from low self-esteem, talk to her about it. If you feel it could lead to depression, consider paying for some therapy for your child.

If your adult child tends to date and bring home suitors more often than you're comfortable with, try these suggestions:

- Because promiscuous behavior can indicate low self-esteem, sit down with your adult child and express your concerns. He may feel like a failure in life and "self-medicate" through a string of relationships. She could have low self-worth and believe she is only appealing to men on a physical level. If you are worried enough about your child's self-esteem, consider paying for counseling.

- If you feel invaded by the changing string of suitors, simply ask your adult child to introduce you to the latest and greatest before spending the night. You may lose track of your son's girlfriend's names, but at least you'll be able to recognize her if you pass each other in the hall going to and from the kitchen in the middle of the night.

It's okay to set guidelines in a situation like this. If you are uncomfortable with something, sit down with your child. Tell your son how you feel. Explain to your daughter how her behavior is affecting you. Ask your son or daughter to do things differently.

What Younger Siblings See

Andrea, a 23-year-old sociology student, moved home during her last year of college. Her younger sister, 17-year-old Morgan, also lived at home, and she idolized her big sister. When Andrea started dating Kevin—and staying over night at his apartment—Morgan just *knew* they were having sex. The girls' parents raised them to abstain from sex before marriage, so Morgan wasn't sure how she felt.

Unbeknownst to them, Andrea and Morgan's parents were sending a mixed signal to the younger sibling. Though it was okay for Andrea to be intimate with her boyfriend, it wasn't okay for Morgan. They never sat Morgan down to explain to her that her older sister was an adult and, therefore, able to make her own decisions about sexual behavior. It seemed like a double standard.

In open-nesting households with younger siblings, it's important that parents sit down with the younger child and discuss why one child is held to a different standard than the other. Explain, for instance, that although you would prefer your children wait for marriage or a very established relationship to have sex, once they've passed a certain age, you simply can't do anything about it. You're giving the older child the freedom, but that freedom doesn't apply to the younger child at this age.

Younger children (grammar-school age, for instance) who don't understand the concept of sexuality can be kept in the dark, but if your adolescent child knows about sex, you can be open with her or him about the situation. Acknowledge that sex is probably happening. Don't try to hide what the older child is doing, because the younger sibling will totally pick up on it, anyway! Discuss what your family values are, and make the entire situation overt and obvious.

Dad or Mom Dating

Single parents in an open-nesting environment not only have to deal with their adult children dating and having sex, but they also have to look at how their own sexual behavior and intimacy issues affect their children. An adult child will be acutely aware of your partners, so it's helpful to understand some of the common concerns faced by single and dating Moms and Dads.

Guilt About Abandoning an Adult Child

Karen, a 45-year-old systems analyst and single mom, let her 21-year-old son, Matt, move back home while he took classes at the local community college. Up until he moved out on his own, Karen had coddled her son. She attended all his rugby games, took him around the world on his summer vacations, and gave him just about everything he wanted. Now that he lived back home, Karen resumed her overprotective and overindulgent routines.

When Karen met the man of her dreams, she felt overwhelmed by guilt. She felt that she was abandoning Matt whenever she spent time with her new boyfriend. Though Matt was an adult, she still felt that she needed be there for her son.

Karen's behavior is typical of single parents. They often have a stronger sense of guilt about their parenting because they feel that their child isn't growing up in the "ideal" family situation. They put more pressure on themselves to be the "perfect parent" and spend more time with their children. Those habits are often okay when a child is growing up, but when the child is an adult, that kind of intense attention is no longer needed. At this point, it's developmentally normal for both parent and child to have independent lives.

If you find yourself feeling guilty about spending time with a significant other, remember that your child doesn't *need* you to be there as much, so you don't need to feel that same sense of responsibility. You can go on that date. You can even stay overnight. If you and your adult child really want to spend time together, consider setting aside time, like a weekly lunch or monthly movie night, when you can be together just for family time.

Privacy, Please!

Andrew, a 55-year-old father, and his girlfriend, Kristy, wanted nothing more than to spend some quality "alone" time together and enjoy a *Godfather* marathon over the long Memorial Day weekend. Unfortunately, Andrew's daughter, Hannah, was a permanent fixture on the living room couch. When she wasn't at work, she was watching HGTV and VH-1 reality shows—not Andrew's and Kristy's favorite thing to do.

Because Hannah rarely left the house—not to mention the living room—after she moved back home, Andrew had little alone time with his new love. In the end, Andrew wound up hooking up a satellite dish to Hannah's bedroom television so he could reclaim his comfy couch—and some alone time with Kristy.

If your adult child's constant presence is preventing you from being alone with your partner, you don't need to resort to banishing your son or daughter. Instead, tell your son that you would prefer to "reserve" certain parts of the house, like the DVD player on Friday night, or the kitchen or dining room for a Tuesday night date night. Communicate openly with your daughter, telling her in advance that you'd like to have some privacy.

You're living in the space together, so both parties will need to give and take a little. For more tips on sharing room, flip back to Chapter 6.

Under Your Roof

Both you and your adult child will want to remember the importance of being discreet. It probably goes without saying, but when you're living together, practice some common sense. You both will be more limited with what you do with your partner, and when and where you do it. Neither one of you would want to get caught doing something that would make the other feel uncomfortable! If your adult child is not keeping private acts private enough, diplomatically bring this up. It is very unlikely that she or he will re-offend!

Your Child Disapproves of Your Having Sex

Sally and her boyfriend, Dave, were a normal, sexually active couple. But Megan, Sally's adult daughter, did not like the idea at all. She disliked it so much, in fact, that she started teasing her mother, asking Sally if she had "fun" last night, and whether she should "give them some space." Dave laughed it off, but Sally felt embarrassed and ashamed by her daughter's sarcasm and taunting.

Depending on your situation, your adult child may act out in a similar way. If that happens, sit down with your child and discuss the fact that you've picked up on the discomfort. Try to act as matter of fact and nondefensive as possible. This may be an uncomfortable topic for your son or daughter (and you, too!) so be as open and honest as possible.

Granted, it may be easier said than done. As we discussed earlier in the chapter, it's typical for families not to talk about sex at all. Sitting down to talk will take courage. It's really uncomfortable and it's really hard, but it's the only way to really resolve these concerns.

Your Child Disapproves of Your Dating

When Bryan met his mom's new boyfriend, Tony, they didn't hit it off so well. Bryan tended to be more laid back and thoughtful like his father, whereas Tony, big and boisterous, had to be the center of attention. Bryan felt Tony was overbearing, and he was concerned that Tony wasn't looking out for his mom's best interests. Frustrated, Bryan decided to talk to his siblings, all of whom lived on their own. They decided to bring the issue to their mom's attention. Mom listened to what her children had to say, and addressed their concerns, but decided to stay with Tony.

Bryan's Mom did exactly the right thing. She listened to her kids, even agreeing that Tony was a little "loud." She acknowledged that no partner would ever replace their father (her ex-husband) to his children. But she was in charge of the relationship, not her kids, so it was her decision to stay or go.

You may find yourself in a situation like this. Remember to keep those lines of communication open. Talk with your child. Address her concerns nondefensively. Thank him for keeping his "eyes open" and your best interests at heart. Do not feel pressured to end a relationship because your adult child doesn't approve. You and only you know what you want in a relationship.

Communicate!

Your adult child may be dating, sleeping over at the partner's house, and likely having sex. Even if you don't agree with it, it's simply a reality in today's world. So how do you handle it? By talking and communicating with your son or daughter.

Keep the conversations with your child open and honest. Don't be embarrassed or ashamed about discussing sex and intimacy. It's absolutely normal for your adult child to be developing those close personal relationships, so expect it, plan for it, and trust that your son or daughter will make the right decisions—undoubtedly after learning from some mistakes.

The Least You Need to Know

◆ Sexual relations are a normal behavior for your adult child at this stage of life. It may be a taboo topic, but it's okay for you to talk with your child about sex and intimacy.

◆ If you're concerned about your child having sex in your home or at a partner's house, communicate your concerns using a relational approach.

◆ Remember that you're adults living together under one roof. If you and your partner—or your child and her partner—want some "alone" time, arrange for it in advance.

◆ Be honest and open with your adult child about sex and intimacy. Remember: having sex is normal!

9

Chores Are Boring, Rent Is Rude

In This Chapter

- ◆ Pitching in with the chores
- ◆ Chore challenges
- ◆ Should you charge rent?
- ◆ When your adult child stops paying rent

Living in a household of adults means sharing adult responsibilities. Just because your son or daughter moves back home, it doesn't mean that life returns to how it was last time your little boy or little girl lived there. Mom is not the live-in cook, maid, and chauffeur, nor is the adult child required to finish chores before seeing friends!

Duties around the house include day-to-day chores, like cooking, cleaning, and yard work, as well as monthly tasks, like budgeting and paying bills. In many ways, the scenario in your open nest should look more like roommates living together rather than a child living with parents. Each family member—including your adult child—has areas of responsibility that contribute both to that person's benefit and to the whole family as well.

Who Does What Around the House?

When deciding who does what around the house, you and your partner should first brainstorm the ideal division of labor in the family before your adult child returns home. Make a list of tasks your daughter can handle (and that you're willing to give up). She'll benefit if she knows what you expect of her *before* she moves in so she knows what she's in for!

When you've decided who's doing what, create a chart (like the following one) to record who will be responsible for what duty. You may wish to list out sub-chores for each of these main categories to make expectations even clearer. (Remember from Chapter 3, goals are more likely to be achieved if you lay them out very specifically!) Family members can share chores, too.

Family Chore Chart

Chore	Who Will Be Responsible?
Taking out the kitchen trash	_____
Emptying the wastebaskets	_____
Trashcans to the curb and back	_____
Vacuuming	_____
Dusting	_____
Straightening up the living areas	_____
Cleaning the bathrooms	_____
Cleaning the kitchen	_____
Cleaning the living room	_____
Cleaning the dining room	_____
Cleaning the bedrooms	_____
Grocery-shopping	_____
Preparing dinner	_____
Doing the dishes	_____
Ironing	_____
Personal laundry	_____

Chore	Who Will Be Responsible?
Household laundry (towels, linens)	_____
Mowing the lawn	_____
Yard work	_____
Add other chores you may think of:	
_____	_____
_____	_____
_____	_____

Challenges with Chores

Despite the brainstorming, planning who does what, and creating a chore chart, issues can sometimes surface. Two common problems that often arise in an open-nesting environment are when parents and child regress into old roles, and when your adult child doesn't live up to his or her end of the bargain.

Regressing into Old Roles

When 23-year-old Janice, a substitute elementary school teacher, moved home with her parents, the family sat down to discuss chores. Janice lived on her own for four years prior, so she was used to doing laundry, cleaning the house, and cooking for herself and her room-mates. Janice and her parents agreed that she would be in charge of cleaning the guest bathroom and vacuuming the common areas once a week, as well as helping in the kitchen when Dad needed her assistance with the cooking.

For the first month, the division of labor worked great. Janice typically substituted three days a week, so on her days off, she completed her chores and still had time to take classes at the local junior college. But things slowly started changing after that. Janice had to study for midterms. She picked up an extra day at the elementary school. Before long, she began rushing through her chores and even skipping them at times. Mom came to the rescue. As family nurturer and caregiver, Mom fell into her old parenting role and took on Janice's chores. Before long, things around the house were back to how they were when Janice was still in high school.

As with Janice and her family, no matter how old we get, we tend to regress to our parenting roles when our adult children visit, just as our grown children tend to regress into their childhood roles. Something deep in our psyche brings us back to that time—after all, we've spent the majority of time with our kids as parents (just as they've spent the majority of time with us as younger children). If this situation happens in your home, try these techniques to get everyone back into their "new" roles:

◆ Use the previous Family Chore Chart to outline your new expectations of who does what around the house. Post it on your refrigerator, if need be, and revisit it at family meetings. If chores need to be swapped or skipped, that's fine—as long as everyone knows about it and is okay with it.

◆ Remain conscious of your tendency to slide back into old responsibilities, which probably means that you—as a parent—are taking on more than you need to. It's important for your child developmentally to assume those tasks and chores.

◆ Let go of any guilt you may have that a "good" mother or father "should" do everything for the child. After all, if your adult child were living alone or with a roommate, 50 to 100 percent of the responsibilities would fall on his shoulders!

Your Child Is Not Doing the Job

Alex, a 20-year-old veterinary technician, was, well, a little lazy. His parents knew it, so when he moved back to Mom and Dad's, they gave him a list of chores to do around the house, like emptying the wastebaskets, washing the dishes, and straightening up the common areas twice a week. He reluctantly agreed.

For the first few weeks, Alex did an okay job. It was nothing like his perfectionist Mom would have done, but he was showing some initiative and responsibility. That didn't last for long. Alex procrastinated on doing his chores. He forgot to empty all the garbage cans. Dishes piled up in the sink. He started out strong, but he petered out as time passed and adopted an air of hopeless resignation.

Under Your Roof

If your adult child resists doing his chores, don't push him. He's more likely to push back if he feels forced. Instead, communicate why you need to have those tasks done. Ask if there's something getting in her way, or if there are other chores she'd rather do. Rather than treating her like a 12-year-old, approach her like a roommate or a friend. You'll get better results.

Making an agreement about the division of labor may be easy, but what about the follow-through? What if you have a lazy son, like Alex? The child may do his chores, but what if the job doesn't meet your expectations? As a parent, chances are you will do your part around the house (okay, maybe not right on time or with 110 percent effort), but what about your adult child who may not be accustomed to (or driven to) doing as much around the house?

Your adult child isn't a kid any more, so you won't want to use discipline like you did when your adult child was a little boy or girl. Besides, he's an adult and should know that lazy behavior has consequences. Instead, investigate *why* your adult child is shirking responsibilities:

◆ Remind your son or daughter about chores left undone, and ask what might be getting in the way of doing them. Tell him how it affects you if his chores are not done. If you and your partner agree that things are so bad that you may want your adult child to move out, tell her how you're feeling. This may add the necessary motivation!

◆ Do a reality check. While talking with your child, you may discover that your son is overwhelmed with his other responsibilities and simply unable to handle taking out the trash. She may be so swamped with her studies that she doesn't have time to mow the lawn. The reason could be deeper than "I just don't want to do my chores."

◆ Explore other deep-seated reasons. You may also discover that your adult child is struggling with unresolved issues, like he's bitter that you asked him to attend a family meeting when he wanted to go to out with friends, or she feels you're being sexist by asking her to help out around the house more than her younger brother. Make time to discuss what might be bothering your child.

Once you've discovered why your adult child isn't doing chores or not doing them well, implement strategies to help guarantee success. For instance, make the chores part of the entire family's routine. While Mom is cooking, Junior vacuums, Dad straightens the family room, and Sally feeds the cats. If your child has trouble staying organized, you can also print out a blank calendar and have your son write in when he plans to do his chores. By doing so, he keeps organized, plus he's accountable for his actions—and his *inactions*. Remember, it's not your responsibility to do your child's chores, but sometimes she needs a little encouragement to get motivated.

Rent? Really?

In our professional and personal experience, few parents charge their adult children rent when they move back into the nest. In fact, we could only come up with one example in which a family charged rent: they felt it would make their child less comfortable being home—and less likely to stay around.

Regardless, you may wish to charge your child rent. It's okay to do so, especially given the extra financial burden another person may add to a household. Here are some other reasons why you might make your son or daughter pay to stay:

♦ You believe that your child needs to learn responsible budgeting to help for success in the "real world." Perhaps your daughter struggled with saving money. Paying rent may teach her to put aside part of her weekly paycheck.

♦ You are concerned that your child needs an "incentive" to leave. And he may take advantage of your hospitality if he stayed rent free; maybe your son is getting too comfortable at Mom and Dad's.

♦ You need the money to cover extra expenses, some of which may be related to a child moving back in. Your child will use water, electricity, gas, the telephone, and food. Expenses can add up!

♦ You are letting your child live in an area of your home that would otherwise generate (much-needed) income, like a mother-in-law unit or apartment above the garage.

♦ You simply think it's "fair" for your child to pitch in financially.

If you do decide to require rent from your child, how much do you charge? This is completely up to you and your partner. You may come up with a flat fee based on the "going rate" in your town for renting a room in someone's home. You may decide on a percentage of your child's income that reflects your area's rent-to-income ratio. You may only charge enough to cover the additional expenses you accrue by your child living at home. Or you may resort to a "made up" amount that seems fair to you, your partner, and your daughter or son.

Whatever you decide, it's important to discuss—and agree on—rent *before* your adult child moves back home. First, discuss with your partner what you think is an acceptable arrangement. Once you've agreed on a plan, propose it to your adult child. Allow your son or daughter to negotiate the terms with you and perhaps reduce rent by taking on some other responsibilities, like house-sitting, yard work, pet care, or childcare for younger siblings. Put your final agreement in writing using a contract. (See Appendix A for a sample contract.)

A New Generation

When you're deciding how much rent to charge your child, a great resource is Craigslist.com. Essentially, the site lists free online classified advertisements from all over the country (and world). You can find out how much a room goes for in your area by first clicking on your state, like Colorado, from the selection on the right of your screen; then selecting the city (or site) closest to you, like Fort Collins/North Colorado; clicking on rooms/shared under the "Housing" section; and scrolling through the results. If you want to get more selective, you can filter your search in the top box by entering a specific city or a particular amenity that most closely describes your home.

When Rent Goes Bad

Charging your adult child rent may work just fine for your family. If your daughter is responsible (or working toward it), the monthly bill may help her on her road to independence. Your son may have to rethink buying that new snowboard—which will teach him how to set priorities. But problems may surface. Here are two common cases of rent wrenching the family's plan.

Not Paying the Rent

Carla, a 23-year-old grocery store clerk, moved home with her folks when she could no longer afford to live on her own. Though she was burdened with credit card debt and a hefty car payment, Carla agreed to pay her parents $250 a month, which essentially covered the added expense of her living with them.

The first month went fine, and when Carla paid her second month's rent 10 days late, her parents let it slide. But when she completely missed her third payment, Mom and Dad sat her down for a discussion. Rather than berate her for being irresponsible, they asked her why she was having trouble paying her rent. It turns out she didn't budget properly, and when the first of the month came around, she had already spent her paycheck by paying her credit card bill, car payment, and insurance, leaving her with nothing for Mom and Dad.

In a case like this, Carla simply needed a little coaching from her parents. They discovered the reason for the missed rent and worked with her on budgeting skills, which ended up resolving the problem. Sometimes your child may have unforeseen expenses, like car repairs or medical bills. Other times, poor budgeting skills may cause your child to miss a payment or two.

Depending on the reason behind the missed rent payment, you may opt for leniency, allowing for a delayed or skipped payment. If you feel your child is taking advantage of your leniency, you may choose to give a stern warning, especially if you think what's needed is to understand "real world" consequences for actions like falling behind or rent or a mortgage. Tell your daughter that she gets one "get out of jail free" card—and that's it. It may cajole her into taking more responsibility, and enable her to feel that you trust her ability to honor her financial commitments. A banker or bankruptcy judge might not be so generous in the "real" world.

But paying rent does have important real-world implications. Your child needs to know that he can't miss rent or go back on a contract in the real world. There are ramifications that can impact a young person's financial

> **Ground Rules**
>
> If your adult child is having trouble managing money, a helpful resource is *The Complete Idiot's Guide to Personal Finance in Your 20s and 30s, Fourth Edition,* (Alpha Books, 2009).

health for years to come. If you're worried that your son or daughter has problems with issues like this, take the opportunity you have in living together under one roof to teach your child how better to handle debt and manage spending.

New Rent Rules

Connie and David decided not to charge rent to their 22-year-old son, Cameron. He decided to take a year off school to refocus his education goals after realizing he didn't like his major subject. So Cameron's parents let him live rent free as long as he agreed to find a job, even a part-time job, as soon as he could, and as long as he agreed to go back to school as planned when the year's time was up.

Three months passed, and Cameron made no progress toward finding a job or deciding on a new course of study at school. He spent his time sleeping in, surfing with his friends, and watching television. Frustrated, Connie and David decided Cameron needed an incentive to get his life moving, so they told Cameron that starting in month four, they'd be charging him rent. Cameron balked and grumbled, but in the end, the strategy worked. He found a job helping out with a web designer, and decided to switch his major from history to graphic design.

Situations like this can be tricky. Luckily for Connie and David, Cameron went along with the change in plans. An adult child who's told *after* moving in that he needs to start paying rent will certainly resist! While it is ideal to make an arrangement about rent *before* your child moves in, you can still do it later on. Reaching an agreement will most likely require a fair amount of negotiation with your adult child, but it can be done—especially if she knows that the alternative entails apartment-hunting.

Take It as It Comes

When it comes to delegating chores and deciding whether to charge rent, every family is different. Look at your unique situation and develop a plan that suits your family's needs.

Remember that during this (brief) open-nesting time of your life, you will be building skills and confidence in your adult child that will prove beneficial for years to come. You're encouraging your daughter to manage her time so that she gets her chores done. You're showing your son how to save a portion of his paycheck to pay for rent or bills. Though it may seem tedious, it's an exercise that will grow your son or daughter—and hopefully get you your nest back sooner!

The Least You Need to Know

- The easiest way for your family to decide on division of labor is to make a list of tasks and assign individuals to each one.

- An adult child who is willing to take on responsibilities around the house may want to participate in brainstorming the list.

- Though few families charge rent to their adult children, it's a reasonable request to make, especially if the child adds significantly to the household bills.

- What you decide to charge will be up to your family. Be sure, however, to get the agreement in writing.

Part 3

Rules? Who Needs Rules?

Do you suspect that opening the nest to your adult child might result in chaos and conflict? Will pandemonium ensue? In this part, we'll explore what it means to live together again. From day-to-day living to sibling rivalries to determining who is in charge, we'll help you figure out the rules of the roost. We've dedicated an entire chapter to meals and food issues, as we know that when, how, and where you all eat is so important to your family's health and well-being.

Chapter 10

Whose House Is This, Anyway?!

In This Chapter

- Making emotional and mental adjustments
- Looking at who makes the decisions in the open nest
- Deciding appropriate emotional closeness and distance
- Finding a happy medium

"If I live in this house, then this is my house, too!" … "A room is not a house." … "When you're under my roof …." Whose house is this, anyway? Whether your adult child has been living in a college dorm room, bunking with a roommate, or in a house of her own, the fact is that your child has already lived away from home and enjoyed a taste of personal space and freedom. It's a delight not easily forgotten!

Now, in moving back home, your child gives up that space for the comfort and stability of your nest. When that happens, old habits or patterns of behavior could resurface, and they may lead to your adult child losing any degree of control or mastery over personal space that he had when living away from home.

In this chapter, we'll explore the concepts of control, ownership, and generosity in a household of adults sharing space. How can you share *your* house with your adult child without ceding control? How can your adult child share *your* house without acquiescing to juvenile powerlessness? Who is in charge?

Making Adjustments

Making physical space in the house is one thing when your adult child moves home. But you'll also want to recognize the mental and emotional adjustments that everyone will need to make, anticipate them, and support your child through them. Here are just some of the things your daughter or son may be going through:

- She may return to her old "role." When she moves back home, your daughter may regress back into being "sullen and silent" because she thinks Dad will still make all the rules or she may rely again on Mom for emotional support. Instead, encourage your daughter to fully embrace her new role as an independent, competent adult.

- He may feel like a failure for not "making it" on his own. Your son could be ashamed about having to return home, especially if his peers are living on their own. You can support him by giving him the tools and the confidence he needs to re-launch.

- She may put life-cycle tasks on hold. Your daughter may feel that it's okay to take a break from pursuing a career path or developing intimate relationships when she returns to Mom and Dad's house. You can help her by encouraging her to set goals centered on her life-cycle tasks.

> **Under Your Roof**
>
> For many young adults, the most difficult aspect of moving back home is the stigma. The American ideal of adulthood is independence—not to mention home ownership—and when a young man or woman must return to the nest, it implies failure. When your adult child moves back in, keep this very real emotion in mind. Support your daughter. Remind your son that it's becoming more commonplace to return home. And share the tools necessary to re-launch, including a plan to help you work together as a family to achieve the goals that will help you *all* succeed as individuals.

Similarly, you may go through some mental and emotional changes, too. When your child moved out, you probably moved on with your life, focusing on friendships, or perhaps volunteering at your local community outreach center. Now that your son or daughter is moving back home, you'll need to make some adjustments, like these:

- ◆ You'll need to learn to recognize your child as an adult. Sure, she will always be your "baby," but to help your daughter get back on her own two feet, resist the temptation to coddle her. You don't have to cook or clean for her, help her get a job, or support her financially.

- ◆ You'll need to set aside your desire to influence your child's behavior and life goals. Instead, be a support structure for your son. Let him make his own decisions. Offer advice (when asked), but keep it at that.

- ◆ You'll need to focus on your own life-cycle tasks. We've said it before, and we'll say it again: keep moving forward with your own personal goals and aspirations. Don't put your life on hold when your daughter or son moves back home.

To sum it up, remember that both you and your child will be experiencing some roller-coaster emotions when you move back in together. Though the changes are temporary, they're real. Be understanding, be loving, and be supportive, and you'll sort out the dynamics over time.

Who's in Charge?

In Chapter 4, we introduced the Circumplex Model, co-authored by David Olson, Ph.D., which is a method for describing and analyzing family dynamics. The model assesses family functioning on the dimensions of flexibility, cohesion, and communication.

Don't worry. We're not going to go through the clinical jargon again! But we are going to put that theory into practice to help you determine how control works in your family using the *flexibility* dimension. If you'll recall, flexibility refers to how much change can take place in the family, how tolerant its leadership is, how the family negotiates, whether roles can be changed, and what kinds of rules it has.

Feelings of worth can flourish only in an atmosphere where individual differences are appreciated, mistakes are tolerated, communication is open, and rules are flexible— the kind of atmosphere that is found in a nurturing family.

—Virginia Satir, American psychotherapist and educator, 1916–1988

Olson identified four levels of flexibility: rigid, which shows little tolerance for change; structured, which is a little more tolerant of change; flexible, which is comfortable with moderate to high levels of change; and chaotic, which has erratic, limited leadership and is continually changing.

Of the four, "flexible" is an ideal family style in an open-nesting environment. The family shares leadership and has a group approach to decision-making. Family members negotiate through conflicts, roles are shared, and change happens when necessary. Parents consider the adult child to be an equal and treat her as such.

Issues surface when a family's style tends to be on either extreme— either too rigid or too chaotic. Those are the families that wind up in Lauren's office. Here's what those extremes can look like.

A Rigid Family

When one person controls the household, it can create feelings of acquiescence and apathy among the family members.

In the Connolly household, Mom ruled with an iron fist. She controlled the budget. She created everyone's schedules and dictated the family activities. She noted everyone's comings and goings in red pen on the master calendar. She told everyone where to go, what to do, and how to do it. For as long as everyone could remember, Mom set the tone and behavior for the household.

As a result of Mom's tyrannical reign over the household, Dad and children, which included 23-year-old Megan, grew to be marginalized from the decision-making process. They just let Mom make all the decisions. Years ago, they learned that it was just *easier* to follow Mom's rules rather than upset her by challenging them. Things have stayed the same ever since.

When Megan moved back home after graduating from college, her mom—not having changed one bit—still treated her just like a child. She told Megan what kind of job to get. She demanded that she be home no later than midnight. Mom wouldn't allow Megan's boyfriend in her room with the door closed. It was just like she thought her daughter had re-enrolled in high school.

Ground Rules

One clear benefit to treating your children as adults is that they can pitch in more around the house. At the same time, give your daughter the opportunity to demonstrate her independence and adulthood, rather than demanding or expecting too much.

Within her family's rigid structure, Megan was once again viewed as a child rather than being seen as a young adult. She was expected to behave as Mom saw fit. Because her life-cycle stage emphasizes coming into her own in her career and relationships, this inflexible environment with a controlling parent was stifling her.

If you (or your partner) tend to control how the family operates (with the best of intentions, of course), it's time to take a step back. It is actually harder on you to put all that effort into keeping track of your adult children. Make things easier on yourself and trust that you've raised a capable adult. It helps your child learn and grow if he can start choosing what he does professionally or in his relationships—as well as managing the minutiae of his day-to-day lives.

A Chaotic Family

Free-spirited may be an exiting way to live your life, but when the family looks like it has attention deficit disorder, that can be tough on all members.

The Jackson household was always in a frenzy of activity—disorganized activity, that is. No one knew what anyone else was doing or where they were going. Mom worked inconsistent hours. Dad would sometimes make dinner, sometimes not. The kids would get themselves to soccer practice or piano lessons—or Mom would take them and they'd show up late because she would forget about an inconveniently scheduled doctor appointment. The Jackson family had no main decision-maker in the family, no process in place to make decisions, and no coordinated communication or organization to speak of.

When Mark, a 25-year-old laid-off graphics designer, moved back into his parents' house, he stepped back into a maelstrom. As a child, that chaotic lifestyle demanded he make his own decisions—even if he wasn't ready—and it made him feel insecure. Now that he was back home, those feelings of insecurity flooded back. Because he was an adult, the craziness didn't bother him as much as it had when he was a boy, but the disarray in the house still made it difficult to know what in the world was going on, which made it difficult for him to make plans of his own.

If your household tends to feel disorganized and like no one is in charge, it's time to institute some family meetings. Pick a time once a week or so for the family to consistently get together to plan. Get out that Family Chore Chart from Chapter 9 and fill in who is responsible for what. Discuss any important issues or decisions that affect the family as a whole. Remember that Mom and Dad are technically in charge (as the Executive System and owners of the house) but that it's okay to operate more democratically, especially with older teens and adult children. Enlisting the help of a family therapist (even for just a few sessions) could do wonders for helping the family environment feel more structured.

Remember: You're in Charge

Sometimes parents don't act like they're "in charge" because they feel guilty about being strict with their children, or they have qualms about showing strong leadership in the family.

For instance, a recently divorced mom felt so guilty about the breakup that she basically dropped all the rules for the children. She didn't want to be hard on them because they had already endured the family's coming apart. Then things started to go beyond the temporary easing up she'd envisioned: she lost her influence with her family. The oldest daughter stepped up and took charge because nobody else was in charge, which caused the other siblings to revolt. In the household, conflict became the norm. Everything fell apart because the parent had ceded control.

As a parent, you don't need to feel guilty about being in charge. It's expected, and it's the most healthful way to be. Of course, with an adult child, you don't have to demonstrate as much leadership as you do with a younger child, but you still shouldn't feel guilty if you have to draw the line sometimes and tell your son or daughter that this is the way things need to be.

Gauging Your Distance

How involved "should" family members be with one another? What's ideal? Again, using the Circumplex Model from Chapter 4, we'll put that clinical theory into practice to help you determine how close your family should be by examining the *cohesion* dimension.

Cohesion refers to the type of emotional attachment that family members have toward one another. Every family has some level of emotional closeness and distance. You may share friendships and interests, make decisions together, and spend time discussing your day with each other. At the other end of the spectrum, you may talk on the phone with your adult child once a month. Each family has its own level of connectedness.

Healthy connectedness refers to an optimal level of emotional involvement between family members. Being *too* connected could lead to the family sequestering itself from the outside world, while not being connected enough could lead to a disjointed family. Your family probably falls somewhere in between.

Olson lists four levels of cohesion: enmeshed, which is where there is too little independence among family members; connected, when family members have a high level of emotional closeness and loyalty to one another; separated, where there is greater emotional separateness; and disengaged, where there is limited attachment and engagement.

Of the four, "separated" could be seen as most ideal between the parents and the adult child in an open-nesting environment. It's when family members share in each other's lives, but time apart is important, too. Some decisions are made together, but individuals generally decide things for themselves. Family members support each other. Some activities are shared, but they generally do things on their own.

Problems arise when families tend to be either too enmeshed or too disengaged. Here, we illustrate what those extremes look like with some case studies.

Family Is Too Enmeshed

Sometimes, family members can be too close for comfort.

Kathy and her single mom Marybeth considered each other "best friends." Or, in current parlance, BFFs. They did everything together—shopping, going to the movies, church, they even went out on double dates together. Kathy called her mom at least 10 times a day, seeking her advice, chatting about the latest office gossip, and sharing secrets. Marybeth's husband left her when Kathy was a baby, so the two women formed a bond—too tight a bond.

When Kathy decided to move back home to save money to purchase a condo, she and her mom spent even more time together. They had breakfast together every morning and dinner together every night. They put little emphasis on relationships outside the family. They even began to mistrust outsiders. One day, when Kathy disagreed with her mom's opinion about who should win on their favorite reality show, *The Bachelor*, they launched into a loud fight that left them both feeling anxious and emotionally drained.

A New Generation

Another manifestation of a family unit that's too enmeshed is when two siblings or a parent and child consider each other "best friends." By doing so, they may be shunning outsiders and becoming too dependent on one another. They often feel hesitant to deviate from what the other person deems "acceptable," and any differences of opinion between them cause high anxiety.

In this enmeshed household, both mom and daughter risk losing out on some important developmental milestones. Outside friendships and romantic relationships, which they should both be fostering, will be limited. Kathy was unable to differentiate herself from her mom, and she became overly dependent on her, even becoming fearful of making her own decisions or forming her own opinions in case she's "wrong." Rather than follow the life path designed for her, she followed the one Marybeth laid out. If their relationship were to continue this way, Kathy's dependence on her mother would hinder her own development.

The best thing to do in a situation like this is to gradually loosen your connection with each other and open yourselves to other people and ideas. If your family is too enmeshed, you should begin to seek friendship and support from people outside the family, relying less on each other. Dive into other life-cycle tasks. Try to increase your tolerance for differences of opinion within the family, too. You are doing your children a developmental favor, and you are not going to "lose" them by increasing your separation—you're actually more likely to hang onto them, because they see they can really be themselves around you.

Family Is Too Disengaged

Sometimes, households can resemble a ghost town. The Perkins family, for instance, had a quiet household. Nobody knew where anybody was or what the others were doing. Dad, Mom, and the children shared very few interests. Mom belonged to a book club and played golf with her friends. Dad went on frequent trips to the Gold Country with his buddies and panned for nuggets. Lisa, 15 years old, stayed in her room and chatted on Facebook with her friends or listened to her music when she wasn't in school or at soccer practice. And Nate, 23 years old, took classes at the local junior college during the day and played in a punk rock band at night. They didn't eat together, share the day's events, or even watch television as a family.

Needless to say, each person was fiercely independent. They didn't turn to each other for support, instead trying to make it on their own. Years ago, Lisa had a fight with her best friend and went to her mom for guidance. Her mother wasn't sure what to say to her, and advised her to "get over" her hurt feelings. Lisa's emotions were squelched and she felt ashamed for expressing herself. That was the last time she turned to her family for help.

Ground Rules

Sometimes, someone in the family needs to set the precedent that it's okay to talk about their problems. Families can easily get into the pattern of not thinking they can discuss certain matters, but if someone in the family speaks up, good things can come from that. Once someone opens the door, she might be surprised by the level of support she receives!

When Nate, still 23, moved back home, he fared just fine in the environment. He grew up with it, so he knew what to expect. Plus, he was smack-dab in the middle of the developmental stage that emphasizes forming romantic relationships and a career path. He would have liked at least some level of involvement from his family, but because they weren't there for him, he went to his buddies for support.

In this situation, the Perkins family would have benefited from a little interdependence within the household, learning how to depend on each other—at least a little! They could organize a regular family event, like a weekly dinner, at which everyone could connect. They could share opinions or discuss a problem or "dilemma" someone may have. They could create some kind of connection with each other.

The Least You Need to Know

- When your adult child moves home, remember the emotional and mental adjustments that will be endured, and support your daughter through them.

- Remember that in your household, you're in charge. Don't feel guilty about setting some guidelines.

- Try to remain as flexible as possible while your child lives at home. If you feel yourself regressing into past roles and starting to control your adult child, keep in mind that your child is an adult and can make his own decisions.

- Give your child emotional and mental space—not to mention physical space—when she's living under your roof. Stay in contact, but try not to encroach on her life.

Chapter 11

Day-to-Day Life Around Here

In This Chapter

- ◆ What to do when conflicts and tensions arise
- ◆ Strategies for when communication breaks down
- ◆ Understanding pursuer-distancer conflicts
- ◆ Logistical strategies for preventing flooding
- ◆ Focusing on the positive
- ◆ When to consider family therapy

You planned for your adult child's stay. You talked as a family and developed realistic expectations for the new living arrangements. You had the best intentions, but conditions "on the ground" proved to be very different from your lofty ambitions. So what happens when things don't happen the way you planned? What if tensions surface, or setbacks slow you down? How do you deal with them?

We've sourced some well-known theorists, including Neil S. Jacobson, Andrew Christensen, and John M. Gottman, who specialize in communication and couples therapy, to shed some light on how to defuse conflict when your adult child moves back home. Much of it centers on behavioral approaches—simple strategies that you can try in your open nest.

In this chapter, we'll explore those techniques for coping with the day-to-day reality of having an adult child living with you. You'll learn how to keep the family on track and how to deal with frustrations, setbacks, and heightened emotions.

Typical Tensions and Conflicts

Not everyone gets along all the time. In the television classic, *All in the Family*, lead character Archie Bunker constantly spars with his live-in son-in-law Michael Stivic about everything from religious and political issues to social and personal issues. Daughter Gloria occasionally gets pulled into controversies, too, resulting in thought-provoking—if not hilarious—debates at the dining room table.

Though the Bunkers represent an extreme example of the types of conflict that can erupt in an open-nesting household, it's more common to contend with day-to-day tensions. Perhaps a daughter slams the door when she comes home at 1:00 in the morning, waking the entire household. Maybe a divorced dad leaves the toilet seat up, frustrating his daughter. Junior could be neglecting his agreed-to chores, even though Mom has asked him time and again take out the garbage or mow the lawn. Many of the issues center on communication (or lack thereof).

There's no denying that conflict and setbacks will occur. It's normal. But when they do happen, are your family's behavioral habits exacerbating the problem? Have your attempts to resolve these day-to-day or even Bunker-style issues led to heated arguments or toward withdrawal (which is especially common in disengaged families)? Don't worry. There are healthy ways to manage conflicts and tension in the household.

> ### Ground Rules
>
> Throughout this chapter, we suggest you use strategies generated by couples therapists to manage conflicts and tension with your adult child. Why? Because a parent and an adult child are on an equal level, just as a couple would be, and the dynamics between the two are very similar. Most family therapy theories assume more of a power differential between the parent and the child because most family therapy takes place with parents and younger (non-adult) children—and that's not really as appropriate in an open-nesting scenario.

Before delving into some common communication setbacks and strategies, we want to remind you about the *relational approach* paradigm referred to throughout the book. When handling conflicts with your adult child, remember to talk to your son or daughter calmly, as a fellow adult, instead of yelling or laying down the law. Keep in mind how cordial (and in control of your emotions) you would be if working out a problem with a friend or colleague. Listen, problem-solve, and negotiate with your child. Any solution you can *agree* on is much more likely to happen than one you *insist* on.

Communication Breakdown

Breakdowns in communication often happen when negative emotions cloud the arguers' vision. Unable to think rationally, they're unable to articulate their concerns and feelings. In an open-nesting situation, the negative feelings could be closer to the surface, especially if a parent tends to lecture Junior like he was a teenager, for instance.

In Chapter 4, we introduced some skills for effectively communicating with your child, including listening, speaking, self-disclosure, and clarity. But what if those lines of communication break down? How do you interact with your child when tensions rise? When feelings heat up? When insults start flying? Following, we've listed some strategies for effectively speaking with and listening to your adult child.

Active Listening

Discussions or disagreements can often spiral out of control, as more and more emotion creeps in and the conversation starts feeling like attack and defense. This often happens because participants don't feel listened to. An approach called *active listening*, as explained by family therapists Neil S. Jacobson and Andrew Christensen in their book *Reconcilable Differences* (see Appendix C), may help to resolve that. Here's how it's done:

1. Gather as a group in a comfortable place. This can be done between two people or even as the entire family. *For instance, Mom and daughter sit down at the dining room table to discuss a recent fight about daughter's long-term boyfriend.*

2. Find a physical object that can be passed from person to person. It can be a ball, a pillow, a pen, or whatever is on hand. *Daughter grabs a remote control from the living room.*

3. The first speaker, Person A, will hold the object and express her feelings, what her point is, what she wants to say. She should limit her statement to several sentences—a "digestible" chunk for the listener to absorb. (She'll have the chance to say more later in the conversation.) *Daughter holds the remote and says that her feelings are hurt when her mother nags her about her "immature" boyfriend.*

4. Person B will then summarize what the first person said. *Mom summarizes that she heard her daughter say that she feels hurt when she criticizes her boyfriend.*

5. If Person A doesn't communicate the point accurately or completely, or if Person B missed something or has misunderstood, Person A clarifies and Person B summarizes again. Person B isn't allowed to say anything until she summarizes what Person A said to her satisfaction. *Daughter adds that she really needs emotional support at this time in her life, and that her boyfriend gives her the support she needs. Her mother's criticism makes her feel like a scolded little girl, incapable of making good decisions. Mom summarizes that her daughter wants more emotional support, which she gets from her boyfriend, and that the criticism makes her feel young and incompetent. Daughter agrees.*

6. Once Person B correctly summarizes or reiterates what Person A said, Person A passes the object to Person B, who can now respond to what Person A has said. The process begins again. *Daughter passes the remote control to Mom, who gets to respond to her daughter's statements. Mom explains why she has been concerned about her daughter's boyfriend and—after her daughter summarizes her statements— asks her daughter what she could do to help her feel more supported.*

This powerful technique slows down the interaction (and prevents it from heating up) and ensures that both people are listened to. The process reveals communication patterns and teaches family members to respect what one another has to say. When you first attempt active listening, it can be challenging to stay disciplined. The first few times your family tries this technique, it may break down and emotions may flare. If this happens to your family, gently remind participants to return to the active listening exercise before tensions get too high. If tempers are too hot to continue, reschedule the conversation for another day and try active listening again. With practice and patience, using this technique will keep heated discussions from spiraling out of control.

Using "I Statements"

"I statements," another communication strategy, have the benefit of softening the startup by sounding less accusing and helping you speak only for yourself. They're phrased as, "I feel [blank] when you do [blank]," or "I feel hurt and unimportant when you walk into the house without greeting me."

"I statements" are based on your *primary emotion.* A primary emotion is usually a very vulnerable one, like feeling hurt, unloved, or unimportant. Often, when people feel those vulnerable primary emotions, they become angry and aggressive because those *secondary emotions* are easier to feel than the primary emotions. Plus, anger is very activating, mobilizing, and protective.

Under Your Roof

Avoid using exaggeration phrases like, "You *always* come home late," or "You *never* do your chores." These hyperboles are rarely (if ever) true, and they will surely trigger a defensive response in the person to whom the statements are directed.

The key with "I statements" is to mine for that primary, vulnerable emotion. Expressing it lowers the listener's defensiveness and creates an empathy response, too. So if you're involved in a conflict with your adult child, use that "I statement" phrasing and communicate to your son the vulnerable emotions you feel. Say, "I feel disrespected and hurt when you leave your dirty dishes on the counter because I'm the one who has to clean them up," rather than, "Why do you insist on contaminating the counters with your dirty plates?" The son certainly won't want to cause Mom to feel hurt!

Be Mindful of "Mind Reading"

As you get to know someone very well—like your adult child—you may start to think you can read her mind or know why she did something. For instance, you may assume that your daughter snapped at you because she's sleepy from being out all night with her friends, but the truth is that she's been spending time at the library applying for college and feels anxious about getting accepted.

Mind reading occurs whenever one person assumes what another person is either feeling or thinking. It short-circuits communication because it avoids the added steps of asking the other person how she feels or what she thinks about something. Because it's almost impossible to accurately assume what your adult child is feeling or thinking, any incorrect assumptions you make could lead to resentment all around.

Make sure to ask your adult child questions (sounding as non-judgmental as you can) before assuming you *know* what's really going on. Remain open-minded, and try to empathize and understand why your adult child is behaving a certain way or making certain choices. Leave the parlor tricks to the experts!

Ask Nicely

When you're requesting something of your adult child, phrase your requests positively. Say what you'd like your son *to* do rather than what *not to* do. By doing this, you're outlining the exact behavior you're looking for, and you're making the listener feel more positive about fulfilling the request.

For example, you would say, "I would like you to call me before you bring your girlfriend over," rather than "Do not bring your girlfriend over unannounced." The former statement is a request; the latter is an accusing demand that could cause a defensive response from your son.

Strike When the Iron's Cold

Sometimes, the best time to bring up an issue is during a period of low-conflict and good rapport. Rather than bring up specific issues during a conflict, do the opposite. Look for a time of calm when you're communicating well with your daughter to bring up the problem. You'll lower the tension level, you'll lower her defensiveness, and you'll generally have greater success. Plus, it will give you more time to think about how to phrase something in a way that your child can receive.

If you're sitting around the living room, for instance, and laughing about your puppy's latest antics, that's the time to bring up something that's bothering you. You could say, "Remember last Tuesday when we were arguing about your workload at school? Well, it bothered me when you said I was being overprotective." This approach is much more effective than yelling at her in the midst of the heated discussion.

Try "Repair Attempts"

If tension is mounting during an argument or discussion, consider using a repair attempt to lighten the mood. Repair attempts, a concept John M. Gottman and Nan Silver refer to in their book, *The Seven Principles for Making Marriage Work* (see Appendix C), are using things like humor, praise, or some other positive cue to defuse or soften disagreements. A repair attempt could be a little joke, a positive comment, or even saying something self-deprecating that momentarily breaks the tension in the conversation and puts everyone back on track.

Acceptance and Change

You *can* change some things about your adult child's attitude, but changing other things may be a lot less realistic. Your son may be forgetful. Your daughter may be a clean freak. Personality traits like these, like it or not, are unlikely to change.

Consider some quirks that your partner may have. Maybe you've been married for 30 years, and you just know—and accept—that your husband, who's a little scatter-brained, is going to leave the bathroom a mess. You've battled and nagged him for three decades, to no avail. At one point, you simply accepted that about him and decided to let it go. You recognized it was something that simply wasn't going to change—and you decided you didn't need to take it personally anymore.

The same can be said for your adult child. As much as you would *like* him to be more organized, punctual, or clean, for instance, if it's not part of his makeup, it's not going to happen. You've raised your son. You've tried to shape and mold his behavior as well as you could, and now that the job is over, it's best to accept him for the way he is rather than continue to try to "parent" or "improve" him. Try to recognize which qualities are unlikely to change, and show acceptance—if not loving appreciation—for them.

A New Generation

According to AARP, we've gone from five million multigenerational households in 2000 to a little over six million in 2008. Clearly, whether out of necessity or not, we're learning to hold our family closer—the arm's-length method of being a family just won't work today. When conflicts and tension arise, families need to work through those issues without tuning out, turning away, or adopting overly negative views. During conversations with your adult child, try some of these strategies. Accept your son's unique quirks, even if they drive you a little crazy. Use positive language, and express how you truly feel about a situation. Ask your child what she thinks and feels rather than trying to read her mind. Communication breakdowns will likely happen, so use these tools to keep the lines open.

Chasing—or Running from—Conflict

Besides communication breakdowns that happen among family members, you may also experience what Gottman refers to as a pursuer-distancer conflict. It's when someone "chases down" the other person to resolve an issue, only to have the other person "distance" himself. It's like a Bizarro version of grade-school tag on the playground. As the pursuer continues to chase, the distancer moves farther away.

The idea is that Person A wants to resolve an issue, so she pursues Person B. Person B withdraws by either physically leaving or emotionally tuning out (stonewalling). The more Person B withdraws, the more Person A pursues. Think of the stereotypical "nagging mother" getting after her son, who responds with "Whatever, Mom …," responding but clearly not *really* responding. The son withdraws, Mom escalates her nagging, hoping an increase in intensity will finally get a reaction, and the cycle gets worse and worse.

A reason for the distancer's withdrawal is often what therapists call "flooding." Flooding, which most often occurs in men but can in women as well, is a physiological reaction that's triggered by emotional conflict. Similar to a person's fight-or-flight response when in physical danger, a person who is flooding may experience the same physical responses, like an increased heart rate, sweaty palms, and inability to clearly concentrate on what's at hand.

For instance, if a bear approaches you in the woods, your body produces adrenaline. Your heart races. Your senses become more acute. Your hands get cold, and you may start breathing faster. Your body is experiencing the natural fight-flight-freeze instinct. With flooding, instead of responding to a potential bear attack, you respond an emotional "attack" from the family member emotionally charging you. When your body does that, your brain switches off, and you're not able to cognitively process what's going on. What you *do* want to do, however, is run away … or, stay and fight.

Distancers withdraw because (often unconsciously) they don't want to get to the point of being flooded. They sense if they stay in the interaction or argument, they would get to that fight-or-flight point and feel very out of control. Many times, they'll withdraw before the flooding actually starts, though they may also withdraw after the flooding happens. But most often, they'll train themselves to avoid the situations that have the potential to trigger that out-of-control feeling.

In an open-nesting household, it's helpful to identify who tends to be a pursuer and who tends to be a distancer. If your adult child is the one who escapes to her room during conflicts, you can be pretty sure that she's a distancer. Or if your son pushes an issue and triggers that fight-or-flight response in you, he's probably a pursuer.

So how do you handle these types of conflicts? Here, we offer some recommendations for avoiding flooding while still fostering healthy discussions.

Take Frequent Breaks

If you're discussing an issue and start to feel that you're becoming flooded, take a break and reschedule the conversation. This should ideally be instigated by the flooded person, who is usually the distancer. If the pursuer reschedules the conversation, she's just doing more pursuing.

Ground Rules

If a person becomes "flooded," he usually requires at least 20 minutes to physiologically and emotionally calm down before he can return to thinking clearly and functioning normally.

For instance, if Dad and Mom are talking to Junior about finding a job and he starts to feel emotionally attacked, he needs to request that they take a break and resume the talk in 30 minutes back in the living room. That way, Junior can calm down, Dad and Mom feel assured that they will be heard, and all parties can think through the debate while away from the tension.

Use Successful Communication Techniques

Both pursuers and distancers will benefit from listening and sharing with each other, especially if they're working through a conflict. The only way to have a productive discussion is if everyone clearly communicates his or her thoughts and feelings in a way that everyone can receive. The distancer has to stop running away and show a willingness to engage in working toward a solution to the problem, while the pursuer has to give the distancer a safe space to talk about what's happening without triggering the old desire to run.

Have you seen the movie *Runaway Bride*? Only when Richard Gere's character gives up pursuit of Julia Roberts's character does she gain the will to look at herself and her behavior and make changes. Once she discovers that she *wants* to be with him, she finds all the motivation she needs to examine her heart, and then to run *to* and not *away from* the man who loves her.

Set Up Weekly Meetings

Weekly meetings offer a wonderful opportunity to discuss issues on a regular basis—and stop issues from being brought up *constantly*, which can make the distancer feel overwhelmed. A distancer-daughter should choose the time for the weekly meeting so that she doesn't feel pursued and experiences a greater sense of control.

Distancers often feel that the other person, her parent in this case, is *always* bringing up issues, negativity, and criticism. With a weekly meeting, the adult child knows that the issues won't be brought up constantly because there's a set time for doing so. The rest of the week, she's free and clear. It really compartmentalizes *when* the issues are brought up, so the distancer doesn't feel ambushed all the time.

Letter Writing

If face-to-face communication is impossible, consider writing a letter to express yourself and give it to your adult child, or vice versa. This last-resort tactic enables the parent or child to pour out his feelings, release them, and share them in a way that's nonconfrontational (and carefully worded). As the relationship heals, the letter writing can eventually lead to physical meetings. If you end up at this point with your adult child, use lots of "I statements" in your letter, use positive phrasing (saying what you would like *to* happen), and try to refrain from accusing your child of anything (which could lead to even more distancing).

Keep It Positive!

Just because conflicts and tension exist (as they do in nearly every family) doesn't mean that your family can't find some positives, too. Look for things that are going well and make note of them. For instance, if your daughter took out the trash and straightened up the kitchen, thank her! Even if she didn't do her other chores, when you praise her you're sending the message that you appreciate what she's done and you'd like her to do *more* of it. It's positive reinforcement at its best.

You may also want to increase the number of positive interactions between you and your adult child. Find enjoyable activities to share with each other. For instance, if you and your son can set aside your arguments when you're at the movies together or on the golf green, do it more! Set up a regular tee-time or plan monthly movie nights. Often, the increase in positive feelings via these activities will squeeze out some of the negativity.

To covertly encourage your child to do something, *model* the behavior you're looking for. If you want your adult child to communicate more with you about her plans, for example, try doing it with her! Call her to let her know when you're coming home. She may appreciate it and start doing the same. Or if you've been looking for an apology from your daughter, you may need to take the lead and show her that you respect her enough to apologize to her.

By keeping a positive attitude, encouraging positive behavior, and pursuing positive interactions, you'd be amazed at how conflict and tension can dissipate. Frame the dialogue with good intent, and the nature of the exchange will feel lighter and the focus will better stay on productive topics. You love your child (and vice versa, of course), so when you're interacting with your son or daughter, praise the positives and you'll see a positive return!

The Least You Need to Know

- Setbacks and conflicts will happen. You'll want to prepare for them by employing communication strategies that encourage speaking and listening skills.

- Most households have at least one "pursuer" and at least one "distancer." When conflicts erupt, a pursuer chases a distancer, which can cause flooding—and halt the discussion. Understand how this dynamic works in your family.

- Keep your interactions positive. Celebrate when things are done *right* rather than focusing on what's *wrong*. Seek out the positive qualities of your relationship with your adult child.

- If you've tried on your own to problem solve but find that your family just can't communicate amicably, consider seeking out a family therapist who can help facilitate healthy dialogue.

Chapter 12

Sisters and Brothers

In This Chapter

- ◆ Understanding the sibling relationship
- ◆ Watching sibling relationships change through time
- ◆ Being prepared for shifting roles and identities
- ◆ How sibling relationships are affected by an open-nesting situation

Author and poet Maya Angelou said it beautifully: "I don't believe an accident of birth makes people sisters or brothers. It makes them siblings, gives them mutuality of parentage. Sisterhood and brotherhood is a condition people have to work at."

With all this talk about parents and adult children, we'd be remiss to neglect the *other* child—your adult child's sibling(s). Every family member deserves a chance to take the spotlight, after all, so we want to focus the lens on the sisters and brothers in the open-nesting family and how they're affected by the temporary change.

The Sibling Relationship

Whether sibling relationships are harmonious or contentious depend on many factors, but in an open-nest household, two factors are significant: age difference and birth order. For parents, the main goal is to encourage siblings—biological, step, or otherwise—to co-exist peacefully in the same space, and understanding how those factors affect relationships can help.

Age Differences

Children close in age, especially if they're the same gender, tend to spend a lot of time together growing up. They play together, share each other's clothes and toys, go to the same schools at the same time, and may have even have the same friends. They grow up during the same stage of their family's life cycle, too. For instance, while Robin, 12, and Kristen, 14, attend to the same junior high school together, Dad and Mom both work full time, and the girls go to Grandma's house after school each day.

While children close in age share many experiences together, these children can also have the greatest competition and rivalry for their parents' care and attention. The younger sibling often has a great need to differentiate from the older sibling to find a niche for herself. In our previous example, if Kristen is a star basketball player, Robin may choose a different way to shine. She may shy away from athletics and immerse herself in jazz flute to distinguish herself from her older sister.

Siblings who were born far apart, on the other hand, often spend less time with each other and have fewer shared experiences. They're interested in different things, since they're in different stages of their lives. Rather than share clothes or toys, the younger sibling gets the hand-me-downs. They have different sets of friends, and they attend different schools. For example, Ryan, 16, attends high school while his sister, Katie, 10, is still in grammar school. Ryan is interested in playing video games with his friends and spending time with his girlfriend, while Katie enjoys horseback riding and sleepovers with her BFFs. They're in totally distinct phases of life.

Brothers and sisters with wide age spacing are still family, but often times they can feel more like only children. They each experience life on their own, though the older sibling may be expected to help care for the younger, and the younger sibling still has her older brother or sister to look up to. But each develops a unique relationship with the parents at a unique time in the family's life cycle. In our previous example, Ryan develops fierce independence because both of his parents worked full time throughout his childhood, whereas Katie develops a closer relationship with her father because he decided to go into early retirement and could spend more time with her during her adolescence.

> **A New Generation**
>
> Because couples are able to have children later in life, remarried couples may choose to have children of their own, even after they've "parented" sons or daughters already. This could create a dramatic age difference between half-siblings, and an older half-sibling could be seen as more of an uncle than a brother.

The narrower the gap, the greater the likelihood is for competition between siblings. The wider the gap, the greater the likelihood is for a disconnected relationship. In an open-nesting household, both have implications, which we'll explore later in this chapter.

Birth Order Differences

In addition to age differences, birth order plays an important role in sibling (biological, step-, and half-siblings alike) dynamics in the open nest, too. Growing up, sisters and brothers have distinct roles and identities in the household that are tied to birth order.

Birth order can influence not only sibling roles in the household, but roles later in life with partners, friends, and colleagues. Let's take a look at general characteristics of each:

◆ **Oldest child**. In general, the oldest child tends to be the overly responsible and conscientious one in the family. Whether female or male, the oldest child can make a good leader, because she has experienced some authority over younger siblings. She can have a serious disposition, and she may believe that she has a mission in life. She may be self-critical, but she often doesn't handle criticism well from others. She may also be protective and conforming.

◆ **Middle child.** In between the oldest and youngest, the middle child has neither the position of "standard bearer" nor the "baby" of the family. A middle child may run the risk of getting lost in the family, especially if all the siblings are of the same sex. A middle child may fall into the scapegoat role, too. On the other hand, a middle child may develop into an effective negotiator, being more even-tempered and mellow than his more driven older siblings and less self-indulgent than the youngest. He may be more creative and non-conforming than his siblings, even relishing his relative invisibility.

◆ **Youngest child.** The baby of the family often has a sense of "specialness" that enables her to pursue her own pleasures without the overburdening sense of responsibility that the oldest child carries. Freed from convention and determined to do things her own way, the youngest child can sometimes make remarkable creative leaps leading to inventions and innovations. But she can also be immature, spoiled, and self-absorbed, and her sense of entitlement may lead to frustration and disappointment. Many parents even encourage the youngest child to remain dependent on them, since they may not be ready to relinquish their parenting roles.

In an open-nesting household, birth order and age differences can affect how each child will react to the adult child moving back home. If you increase your understanding of your children (and what's normal and expected), you will have more patience and you'll know when (and if) it's time to intervene in your children's dynamics.

Sibling Relationships Through Time

Now that you understand how age differences and birth order affect sibling dynamics, we also want to introduce how siblings' relationships with each other change over time. It's another variable that will impact the open nest, especially if the siblings are in different life stages.

As Adolescents

Growing up, siblings go through all sorts of physical, emotional, and mental changes. As adolescents (ages 13 to 21 or so), children look for their own identities. They're increasing their understanding of themselves in relation to their peers, family, and community. And they're developing the ability to work alone and with others in constructive ways. They're in those teenage years when just about anything can (and does) happen!

In adolescence, siblings become important models for one another. An older brother, for instance, may head down a creative path, whereas the younger brother takes an academic path, but they both vicariously experience one another's journey. Sisters in particular tend to share secrets, clothes, and sensitivities about their parents' problems.

This stage is also when many childhood rivalries and hurts begin, which may carry over into early adulthood and beyond. Because adolescents are forming their philosophies for life, emotional competencies, abilities to self-manage, and moral identities, they can be quite hurtful to one another and this can affect their sibling relationship for years to come. Not all sisters and brothers are close, and adolescence is when seeds of distance can be planted.

Under Your Roof

Parents may inadvertently perpetuate old sibling patterns. A mother may compare one child to another, or a father may talk repeatedly of how proud he is of his daughter, not realizing he's ignoring his son. A parent may show support of one sibling to "shape up" another. It's a practice to be aware of, especially if your children are displaying signs of jealousy toward one another.

As Young Adults

As young adults (when they'd likely be in an open-nesting household), siblings may feel closer than ever before, especially just before they launch from the nest. They're getting older, looking at the big world out there, and they're recognizing that they'll be on their own soon—and that they might even miss each other!

Once launched, siblings then go through a period of distancing, where they're developing new relationships and friendships, and establishing careers and families of their own. They may get together during holidays at their parents' home, but often their focus is on their own social lives or families or, sometimes, their parents.

Here are some hallmarks of this stage in development:

◆ **Minimal support.** Because they're focused on their future, siblings often show minimal support for one another. They distance themselves to establish their own identities independent from their family or siblings. An older brother building a career, for instance, may not have the time (or inclination) to rush to his little sister's aid at the drop of a hat.

◆ **More competitive.** At this time in their lives, competition between siblings may be the strongest. They may compare successes of their spouses, or they may try to best each other in their careers. They're proving that on their own, they can be successful—more successful than their brother or sister.

◆ **Memories from childhood come back.** Images that each sibling develops of the other may be colored by rivalries carried over from childhood. Oldest sister may still resent her "spoiled and irresponsible" little brother. Or a younger sister who felt dominated by her older brother may feel uncomfortable sitting at the same table with him.

Depending on their phase in young adulthood, your children may be closer emotionally or more distant as siblings. The implications for the open-nesting situation can range from beneficial, when they could be supportive of one another through the rough spot in life, to detrimental, when they're competing with one another or struggling with old sibling rivalries.

Sharing Resources and Shifting Roles

When an adult child leaves the nest, the younger siblings enjoy a windfall of family resources, like parental time, attention, and praise (that is, appreciation of their good qualities, achievements, and interests); money; space in the house; cars; and electronics. When an older sibling moves out, the younger siblings usually get a bigger share of these resources, which can be a big plus for them. A younger sibling may suddenly see Mom and Dad at all of her soccer games; get sole use of the "kid's car," X-Box, and media room; and receive more expensive birthday presents.

When the older sibling returns home, there will be a re-shifting of resources and roles, as well as personal identities. Family members tend to regress to their old roles and identities (since it's what they know best), but some siblings may feel resentful or resist it.

Any "addition or subtraction" of family members causes stress to the family, even if it occurs under positive circumstances. A younger sister who blossomed in her sister's absence may once again feel she's back in her older sister's shadow. She may receive less parental attention and feel less competent. She may experience sadness, frustration, anger, or other difficult emotions in reaction to this re-shifting.

If she had a good relationship with her sister, she may also enjoy many aspects of the sister's return. They may support each other through difficulties, enjoy each other's sense of humor, and split chores. The older sister may offer rides or useful advice.

Siblings in the Open Nest

In an open-nesting household, the dynamics we've talked about in this chapter can affect how the siblings react to one another when the adult child moves back home. Remember, too, that they may also be shifting roles and identities. For instance, the youngest child, who bumped up to only child when her brother moved out, is now back in the youngest-child role.

A younger brother may see the opportunity to get dating advice from his big sister, or a younger sister may ask her older brother to drive her to school instead of their mother. Though tensions and rivalries may exist, it is possible to come back together as a loving family and support each other up until the adult child re-launches.

In the following sections we discuss some of the most common situations that can occur between siblings in an open-nesting household, and how you as parents can manage them so that everyone benefits.

Between Launched Siblings

If all of your children have already launched and one child needs to return home, the independent child may feel jealous over the attention, time, or money the open-nest sibling may be receiving. He may feel that it's not fair that the other child is seemingly "rewarded" for being unsuccessful. This can be especially problematic between brothers, who are already competitive, or between an oldest and youngest sibling, where the youngest moves back home to be further "babied."

When Darin, 22, moved back home, his older sister, Naomi, 28, felt resentful. Darin had always been the baby of the family—and that allowed him to get away with all kinds of childish behavior, like being lazy and unmotivated. Darin was a brilliant woodworker, but rather than find a job using his skills, he just cruised through life, and Naomi felt that their parents were enabling his vagrant lifestyle.

Rather than talk to her parents about how she felt, Naomi just stopped going by her parent's house. She had a close relationship with her dad and mom, but she couldn't stand seeing Darin laze around the house and mooch off their generosity. Ultimately, this caused a rift in the family. The siblings stopped speaking to each other, even after their mom tried to restore their relationship.

In this situation, the siblings' birth order, gender, and age difference all came into play. Naomi, the oldest, resented the dependence Darin had on their parents. Being a woman, she felt compelled to right the situation but she didn't know how. And being six years apart, the two hadn't formed a very strong bond.

When Naomi refused to go to her parents' house, the siblings' parents could have first addressed the issue. In it, Darin could have explained why he needed to move home, giving Naomi a greater understanding of his predicament. Naomi could have gotten her feelings of resentment off her chest and could begin coming to terms with them, perhaps even with her family's help. If handled correctly, the experience could have helped to create that bond the siblings hadn't established before, while growing up.

One Home, One Launched—and Back

If a younger sibling still lives at home and an adult child moves back, a range of scenarios could surface depending on the relationship between the siblings. As we discussed earlier in the chapter, siblings will always shift roles and identities. Here are some other issues that may arise:

◆ The younger child will need to shift roles again. When the adult child moved out, the younger sibling became an "only child" in the household. She'll now need to adjust and become "younger sister" again—and compete for Mom and Dad's attention.

◆ The younger child may feel that he needs to fight for space in the house again, too. Maybe the siblings will need to share a bedroom or bathroom again, or they'll need to share video games, televisions, or movies, not to mention common space in the house.

◆ Old sibling rivalries may resurface, too. It used to drive older sister crazy when her little brother left his dirty socks in the bathroom, so she would yell at him. Or the two brothers constantly competed for Mom's attention, and even though they're older, they find themselves outdoing each other to get Mom's praise.

◆ If the siblings got along well their entire lives, an open nesting situation could strengthen their relationship even more. Two sisters could discover they both love to cook together, or two brothers may find they love to play guitar and decide to form a garage band.

It's important for parents to keep the younger siblings in mind when the adult child moves back home. Try to empathize with your son's or daughter's situation, and do what you can to minimize the impact of the adult child's return. Include them in family meetings, and encourage them to speak out about what they need to be happy in the new living situation.

Under Your Roof

If the younger sibling shows signs of sadness or anger upon your adult child's arrival, be sure to address those negative feelings early on. An adolescent girl may feel that her parents are neglecting her when her older sibling moves back home, and that could cause her to act out or withdraw. A young man may feel anger toward his sibling and not know how to constructively direct it. This transition time can be a golden opportunity to foster open dialogue in the family. Remember, open communication of feelings within the family is a key for healthy differentiation.

Between Step- or Half-Siblings

Blended families are common in the United States. Whether it is because of the divorce and remarriage rate, or single parents having children and getting married later, many brothers and sisters come together with one shared parent. This dynamic can cause a host of emotions and issues between not only the siblings, but the parents, too.

When an adult child moves back home, the siblings from the step-parent's side may feel resentment, that the adult child is taking resources they feel they're entitled to, or that the parents are playing favorites. As with other scenarios, it's critical for parents to communicate with their children openly and honestly. If tempers flare, you may consider finding a family therapist to help your family come together and function in a healthy way.

A disagreement could erupt between parents, too, if one partner's child needs to move back home. As we've discussed elsewhere in the book, a step-parent may not feel comfortable footing the bill for his partner's child if he feels that she's enabling the negative behavior. If this happens in your open nest, talk to each other openly and come to a compromise of some kind. Don't be afraid to seek consultation from a family therapist, as preserving your marriage is top priority. It can be worked out, as long as everyone has the best interests of the child at heart.

The Least You Need to Know

- ◆ Siblings' relationships are affected by age differences and their birth order.
- ◆ The relationships between brothers and sisters changes through time, depending on where they are in their life cycles.
- ◆ Siblings will experience shifting in roles and identity when the adult child moves back to the nest.
- ◆ When an adult child moves back home, his siblings will react in a variety of ways. It's critical for parents to be prepared for their children to feel jealousy and resentment.

Chapter 13

What's for Dinner?

In This Chapter

- ◆ Who buys the food?
- ◆ Planning and preparing meals as a family
- ◆ Sitting down together (or not)
- ◆ Resolving common food-related issues

We mean it: what's for dinner? Answering this question reveals so much about your family dynamics. Who is shopping (and paying)? Who is cooking? And what does everybody want? Learning to plan the meals and cook for the family takes focus and discipline—especially in a household of adults with different tastes and schedules.

How your family interacts around the kitchen table also speaks volumes about your family dynamics. Does everyone eat the same meal at the same time, or does each person do his or her own thing, eating different foods at different times? Do you share conversations about your day, or do you watch television, listen to your iPod, Twitter your followers, or check out your Facebook friends in between bites?

You're Buying!

Part of you and your child's open-nesting plan (and contract, if need be) should include something about who will be buying the food. It's important to discuss this point, because if you're not financially prepared—or just not willing—to feed another mouth again, those snacks and incidentals can add up and you may need to ask your child to pitch in. According to the U.S. Bureau of Labor Statistics, the difference could be more than $100 a month—which is a lot of money these days.

A New Generation

In 2006, American households with adult children spent an average of $5,578 per year on food at home—that's $465 a month. Households with *just* a couple spent $3,571 a year, or $298 a month, according the U.S. Bureau of Labor Statistics.

In most cases, you, the parent, will be stocking the shelves with staples, like tomato sauce, pasta, rice, peanut butter, milk, cheese—the things you buy and use regularly, regardless of whether your child (and maybe his or her own family) is living with you. But when it comes to *special* foods, like your daughter's favorite Ben & Jerry's ice cream or your son's two-bag-a-night tater-tot habit, are you willing to foot the bill?

Talk to your partner about it. Make sure you're thinking about your own future (and retirement); if you can afford to fund your adult child's gastronomical cravings, fund away! But if it's going to deplete your bank account, it's okay to ask your child to shop for her own ice cream or his own tater tots.

Menu Planning and Preparation

If your family has decided to share meals together, planning the menu in an open-nesting environment can be a fun experience for everyone. Rather than leaving it to Dad or Mom to come up with new and delicious recipes, sit down as a family and brainstorm some creative menus. Here are some ideas to get you started:

◆ Have each person bring five of his favorite recipes to the table, put them in a hat, and choose some for the week.

◆ Come up with "theme" nights, like fish taco Friday, pizza night, or Szechwan barbeque night.

◆ If you're trying to lose weight or incorporate healthy eating in your diets, have everyone come up with her favorite chopped salad or creative way to include veggies in the menu.

◆ Scour foodie magazines, like *Bon Appetit,* go online at the local library to download new recipes, or watch the Food Channel for novel recipe ideas.

Depending on your household, you may decide to create a regular meal rotation, or you may choose to change the menu weekly. By tossing ideas around as a family, you'll come up with a diet that will suit everyone's tastes.

Once your menu is planned, who will be responsible for grocery shopping? Will you do it as a family? Or will Dad do it on his way home from work? How you handle shopping duties will really depend on your household. It can be easier for some families to have one person in charge of shopping. For others, it's more convenient for different family members to stop at the store on the way home from work. Discuss it among yourselves to come up with a plan that works best for your open nest.

Ground Rules

In a busy open-nesting household, a dry-erase board can do wonders for consolidating everyone's food requests in one convenient place. The boards come in all shapes and sizes. Tack one onto your refrigerator and request that everyone write down his specific cravings.

Is someone in your family a budding gourmet chef? Does one person "tolerate" cooking duties? Or does everyone dread the idea of even picking up a saucepan, let alone making a mushroom marinara? As more families trade dining out for dining in, that means one thing: someone has to cook the meals, and someone (eventually) has to clean up the mess.

When your adult child moved back home, did you assume that meals would return to the way they were before he left? Schedules and priorities have probably changed. When deciding who will play chef and how cooking duties will rotate in your open-nesting household, consider these questions as a family:

- Will the designated chef be preparing all three meals for each family member? In most all-adult households, this is simply impractical. An alternative (as each person *is* an adult, after all) is to have each person prepare his own breakfast and lunch, leaving only dinner to be made for the whole family.

- Will one person take on the cooking duties? Is that budding gourmet chef inside Dad itching to try a new recipe every night? Does Junior find it relaxing to cook—and so doesn't mind spending time in the kitchen every night—as long as someone else will do the clean-up? Does Mom want to revisit her old "motherly" duties? If so, that may work just fine for your household.

- Would each person be willing to commit to a night (or two) of kitchen duty? If your family tends to be the "tolerant-of-cooking" variety, perhaps a different person could cook each night. Create a schedule, and let each person choose and cook the meal on her "night." For instance, Jane will cook her famous twice-baked potatoes on Monday, Dad will barbeque tri-tip and grill some zucchini on Tuesday, and Mom will toss together a big Greek salad on Wednesday.

- Are once-a-week family dinners more your style? The schedules in a bustling household of adults may not allow for sit-down dinners every night, so plan a weekly Sunday dinner or Friday family movie night when everyone pitches in to cook and enjoy a meal together.

Regardless of how your family assigns cooking duties, someone will need to clean up the mess! Back in Chapter 9, we discussed chores in an open-nesting household. You may choose to include "kitchen clean-up" in your adult child's list of chores, or you may come up with an agreement that when Dad cooks, Mom does the dishes, or when Mom cooks, Junior cleans the dishes. As a family, decide how "clean-up duty" will work best for your situation.

How Is It Enjoyed?

Eating together at the dinner table looks nothing like it once did. In an age of iPods, BlackBerries, relaxed etiquette, and multitasking, your children's table manners can mean listening to an iPod with one ear and Dad and Mom with the other, gulping dinner, and texting friends at the same time (all with a television show playing on the screen of a laptop propped on an extra kitchen chair)—and doing it all quite well.

"Hip" parents may be totally into it. They may have their own Facebook profiles, constantly updating their "status" as the day passes. They may "tweet" all day, commenting on what their friends are doing on Twitter. Many parents (and grandparents, for that matter) have cell phones or BlackBerries or iPhones of their own, not to mention e-mail accounts, websites, and blogs. We're all plugged in—but we're not necessarily on the same page when it comes to etiquette. Whereas young adults may consider it cool to text during dinner, for instance, parents may think it's downright rude.

Your child is an adult, and she is old enough to make her own decisions about who, when, and how much to text or reveal through social networking sites. But remember that if your feelings are hurt when your child tunes you out, it's completely reasonable for you to make requests for how technology is used (or not) while dining together.

Resolving Common Issues

When it comes to family dynamics around an open-nesting dinner table, you can expect that issues will surface. They may center on *what* to eat, *how* the family eats together, and *who* does the cooking and cleaning. Here, we've listed a few of the top conflicts that may come up and how to resolve them as a family:

- ◆ **Communicate your expectations.** Food preparation in your open-nest household could be quite different than it was when your child first flew the coop. Dad may have prepared breakfast, lunch, and dinner for the family years ago, but now he may be too busy building his catering business to cook for the family. If this is the case in your household, be sure to talk to your adult child about expectations for family meals to develop a strategy that works best for everyone.

◆ **Deal with disappearing food.** You plan the meals for the week and do the grocery shopping, but when you go to cook, the food is missing. What do you do? Make it a rule that if someone eats all of something, he either needs to replace it or write it on the grocery list. If someone knows he *needs* something for an upcoming meal, he should put a sticky-note on the item saying "hands off!" or ask the family members to leave it be—no matter how tempting.

◆ **Manage mixed dietary requirements.** In a household of adults, different people have different tastes—and possibly dietary requirements—that everyone needs to be aware of. Rather than seeing it as a burden, see how you can embrace the new way of eating as a family, thereby encouraging and supporting someone you love at the same time, and helping to ensure that loved one's success … and perhaps even survival to a ripe old age.

The Least You Need to Know

◆ Before your adult child moves back into the nest, decide who will be paying for groceries. Will you purchase the staples while your adult child pays for her own treats?

◆ Decide early on who will be the household chef and clean-up person. It may be one person, it may be a group effort, or it may be that everyone's on his or her own.

◆ It's your home, so it's reasonable for you to make requests about how technology is used (or not used) when your family sits down for a meal.

◆ Resolve common food-related issues as a family by keeping the lines of communication open.

Part 4

What, Me Worry?

Are you concerned that opening your nest to your adult daughter or son will mean that they'll be bringing a lot of potentially challenging problems or situations home with them? In this part, we'll explore some of the immediate subjects of worry that might spring to mind ... such as how to handle money and financial entanglements, or what happens when your adult child moves back to the nest and brings *his or her* children (your grandchildren!) along with them. You'll learn strategies for using humor and positive thinking to encourage tolerance and acceptance.

Chapter 14

Holding On to Your Money

In This Chapter

- ◆ How well do you manage your money?
- ◆ How well does your adult child handle her money?
- ◆ Understanding how money fits into the open-nesting environment
- ◆ Exploring the key issues surrounding money

Financial worries and problems top the list of reasons why adult children move back home with their parents, so it's completely understandable for you to be concerned about protecting your own financial well-being. If you're like most parents, you want to help your child, but you don't want to do so at the expense of your own retirement savings.

Is there a way to do both? Can you expect to financially support your son or daughter—for a short while, at least—and still have money in the bank later on? Yes, it *is* possible, but the solution involves teaching your children the tools to manage their own finances.

Money Management 101

Who taught you how to manage your money? Did you stumble through your early adulthood finances, learning about the pitfalls of credit or the importance of paying your bills on time? Were your first savings lessons all about putting spare change in that piggy bank—and then they stopped there?

While this current generation of young people have to prove themselves to a bank to earn the privilege of a credit card or loan, they're more dependent on easy credit to leverage their launch than were previous generations. As a result, many are drowning in debt and have little money in savings. It's not a fun spot to be in … and especially not in these times, when "easy" credit is not so easy to come by anymore.

Retirement Ahead—Ready or Not!

You know you worked hard for your money. You've been reading a best-seller on finances or investing here and there over the years. You've heard of Suze Orman and possibly Paul Krugman. Maybe you've even been known occasionally to read *The Wall Street Journal*. Many soon-to-be or current retirees research financial matters and begin to invest or save in earnest, and many have worries about health-care costs, the economy, and plunging home values.

The truth is that in recent years the majority of Americans are poor savers. Our growing consumer mentality and reliance upon credit has shifted our focus from saving to spending. In fact, according to the U.S. Department of Commerce, Bureau of Economic Analysis, the personal savings rate has steadily declined since 1984, when 10.8 percent of disposable income went into savings compared to 0.4 percent in 2005, and rising slightly to 1.7 percent in 2008, thanks to the desire for self-protection brought on by the current recession.

Statistics aside, today's economic situation for retirees feels more than a little shaky. Will *you* have enough money to retire? If you're not sure, you may want to rethink giving too much cash to your adult child. Now is the time to be very conservative with money, because nobody knows what the future holds. Make sure you're secure in your finances, because the last thing you want to do in 10, 15, or 20 years is borrow money from your child!

A New Generation
The best savers in American history lived in the early 1940s, during the Roosevelt presidency. According to the U.S. Department of Commerce, Bureau of Economic Analysis, the savings rates were as follows: 1941, 12 percent; 1942, 24 percent; 1943, 25 percent; 1944, 26 percent; and 1945, 20 percent. From there, it remained a steady 8 percent or so until the early 1990s, when it plunged precipitously, from almost 8 percent in 1992 to 0.4 percent in 2005.

Exorbitant Expectations

Despite the financial reality that many households face today, Americans remain a hopeful bunch, and many hold on to the belief that they'll come into more money in the future. The stock market may surge. Their home's value will once again skyrocket. The American way centers on hope for the future, right?

Teenagers today have an optimistic look at the future—perhaps too optimistic. According to Charles Schwab Teens & Money 2007 Survey findings, 73 percent believe they'll be earning "plenty of money" when they're out on their own. Fifty-three percent believe that they will do better than their parents or guardians. And, incredibly, teens believe they'll be earning an average annual salary of $145,500 (boys expect to earn $173,000; girls $114,200).

Many adults, young and old, are quite unrealistic about their finances. Retirees may not have saved enough to get them through their Golden Years. Young people may start "consumption smoothing," where they predict that they'll make more money later on, thereby justifying lavish lifestyles and out-of-control spending with the hope that they'll be able to pay for it down the road. These out-of-reach ways of thinking lead many into a cycle of debt.

Money in the Family

Most people associate money with security, freedom, status, and happiness (well, the cultural belief that "things money can buy" can make you happy, thanks to the heavy influence of advertisers). People feel very emotional about money, and in the household, finances are the number one issue couples argue about and seek counseling for.

Let's take a closer look at how your personal history affects how you approach the almighty dollar.

Family History

How you spend and save money often mirrors your family of origin's ways of spending and saving. Most commonly, people tend to do exactly what their parents did. For instance, if your parents clipped coupons and kept to a strict budget, you likely did the same for your family, too. Other people, however, may overcompensate in the opposite direction. When two people come together, they bring two different family-of-origin money-management styles to the table. It can be challenging to meld the two and agree how the household will be run.

Money may also cause a power differential in the family. There's often the perception that whoever earns more money has more say in the big decisions. The lesser-paid partner (or even adult child) may assume that the top wage earner should have the final say on whether the family buys that house or takes on an additional car payment. It's not necessarily true, but that *perception* of inequality can create an uncomfortable tension in the family, not to mention low self-esteem in the person who earns less money. A stay-at-home partner contributes home and child-care duties that have a tangible value—and a precious intangible one as well.

Teach Them Well

No matter how you and your partner handle the finances, it's critical that you teach your adult child solid money-management techniques. You want your son to understand the dangers of credit card debt. You want your daughter to grasp the importance of saving her money. By instructing your child about finance basics, coaching through the difficulties, and helping him experience how money works, your child will be far better off.

When your adult child moves back home, take the opportunity to turn money stress into financial confidence. If your child is open to learning, schedule an appointment for him with your financial adviser for some guidance. Invest in a money-management basics book, like *The Complete*

Idiot's Guide to Personal Finance in Your 20s & 30s, Fourth Edition (see Appendix C). Show your daughter your IRA and how it's grown (or shrunk) over the past year. Invite your child to participate in the household money routine. Give your child the tools and skills to become financially literate.

Finance Issues in the Open Nest

Money is an issue in just about every household, including one with adult children. As we mentioned earlier in the chapter, money is the number one issue couples argue about. When your child moves back home, you and your partner will want to consider these common issues that surface and prepare for them.

Bailing Them Out

Parents should be careful not to let fear cloud their vision—and drain their life savings. Recently laid off, Morgan struggled to pay his mounting credit card debt, student loans, and mortgage. He fell three months behind and the bank had started to send him foreclosure notices. He couldn't find a job and was on the verge of declaring bankruptcy at the young age of 26.

Morgan's parents, Jim and Alisha, watched helplessly as their son's finances dwindled. They invited him to move back home, but they wanted nothing more than to bail him out and see him keep his home. They felt anxious about their son's future and the long-term ramifications of bankruptcy. It would ruin his credit score! They decided to take on his mortgage payments until he got back on his feet again.

In this case, Jim and Alisha feared Morgan getting into serious financial trouble, and they put their own financial well-being at risk. It's understandable that they felt anxious about Morgan's future. But rather than immediately bail him out, they should have helped him look for solutions to his problems.

Money problems do cause anxiety and outright fear, for both the parents and the adult child. If you find yourself in this scenario, try these techniques:

♦ Try to calm down and think level-headedly about the issue. Examine worst-case scenarios with your partner and your child. What if your son does lose his condo? What if his credit rating is awful for the next seven years? What will he do? Is there a way that he will get through it?

♦ Problem-solve ways for your son to address the problem himself. Doing so will be an important learning experience for him.

♦ Examine the trade-offs of helping or not helping. Create a benefit-drawback list. Discuss them with your partner and come to a sensible solution—and one that doesn't involve spending your life savings. If he truly has no way out, examine the financial consequences of *you* bailing him out.

Ground Rules

If you're worried about your adult child's finances, as part of your problem-solving exercise, try to consider the experiences your child has had before. Did he successfully pay off his credit card debt? Was he able to work with lenders? If so, have confidence that he will be able to resolve the issue again. Of course, the opposite could be true, too. If your son flaked on paying his bills two years ago and he has collections agencies calling again, he may need a little more help—and some financial counseling from a professional.

Remember to lean toward any solution that involves your adult child solving the problem himself, which will teach him that he *can* get through tough situations, and that he's responsible for his own decisions.

Enabling Entitlement

An adult child may feel that she's entitled to financial help, which causes her parents to feel guilt-tripped into helping—and enabling—those poor financial habits.

Nadine attended a pricy private university, and she was surrounded by peers whose wealthy parents gave them everything they wanted—cars, clothing, trips to the Bahamas, even their own condos. When she burned through her financial aid fund and couldn't afford her rent any longer, she moved back home with Dad and Mom, and back to her middle-class lifestyle.

Almost every day, Nadine told her parents, Bob and Gloria, about her friends' upscale apartments and condos, and how great it was that they lived on their own but didn't have to pay a single dime. Feeling that they failed their daughter by not giving her an apartment of her own, Bob and Gloria decided to take a second mortgage on their home and "invest" in a condo for Nadine while she attended school.

This scenario illustrates the power of the guilt trip. Nadine, burdened with low self-esteem, spouted on and on about how well her peers had it. Bob and Gloria both came from lower-middle-class families of origin, and they knew how it felt to not have what their friends had. Plus, they wanted to have a friendlier, more understanding, and more hands-on relationship with Nadine, so they put their own financial security aside to give their child what she thought she wanted.

> **Under Your Roof**
>
> A child's sense of entitlement is often not about laziness or greed, but rather about low self-esteem or a lack of skills to earn what she wants herself.

Does this situation sound familiar? Does your child have a sense of entitlement? If so, consider these suggestions:

◆ Examine what's behind your daughter's apparent sense of entitlement. Having benefited from significant parental support, she may feel she isn't capable of taking care of herself (low confidence), or that it isn't safe to take risks. She may lack necessary life skills, or your child may have a mental illness, like depression or substance abuse, that keeps her depending on others.

◆ Based on your assessment of what's going on, problem-solve with your spouse and adult child. How can you most effectively address the problem (again, erring on the side of having your child take on *more* of the responsibility)? Does your child need skills, a confidence boost (best achieved by doing something for herself), or mental health treatment? Helping your child gain these things is *not* enabling. It's teaching your child how to do it on her own.

◆ Pull back on your own guilt. If you are helping your child get to the bottom of what is preventing her from being financially independent, you're doing all you need to. Your child will still feel loved—and love you back—if you stop financially supporting her. Remember: you don't need to buy your child's love.

These problem-solving techniques can also be used if your adult child misuses her money, either her own or funds provided by you. If she doesn't save her money, if she buys beer or drugs or shoes, or if she lets bills fall by the wayside, examine what could be stopping your daughter from spending money responsibly. Form a plan for addressing the problem. And as soon as it's practical, stop contributing money to your adult child (at least until she demonstrates responsible use of it).

The opportunity to draw support from being home again can provide much more tangible comfort for your child than any dollar bill can give. When you open your nest, you can reinforce the value of family relationships and commitments, and your adult child can see the benefit of shared family purpose. As many a grandmother raised during the Depression preached like a mantra during our own childhood years: "We didn't have anything but each other … and that was all we needed." When you set goals as a family and work toward them responsibly, and together, everyone benefits.

Couple Tension

You may disagree with your partner about how much support to give your adult child, especially if you're divorced and remarried. David and Cherie, for instance, had been married for two years when David's son, 22-year-old Troy, moved back home. With no real direction in life, Troy had spent the past year couch-surfing across the country, staying at friends' houses from California to North Carolina. He ran out of money (and friends' couches) and had to return to his dad and step-mom's house and find a job.

David agreed to let Troy live with them for as long as he needed, and told him he didn't need to sign any sort of contract. That bothered Cherie. She loved David and Troy, but she didn't feel comfortable supporting the young man financially—especially after his track record. Troy already owed David $10,000 for paying off his credit card bills. Cherie felt David was enabling his son's vagabond tendencies.

In a situation like this, it won't take long for the tension in the household to mount. David strongly believed that he should take care of his son, and Cherie strongly believed that Troy should act like and adult and learn to take care of his own financial worries. If this situation rings true in your household, remember that as a married couple you are the "executive system," responsible for reaching decisions together about matters like money. If you cannot reach an agreement on how much to help your child, seek out a couples or family therapist to help you. Do not take matters into your own hands and undermine your spouse.

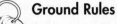

Ground Rules

Form a united front with your partner when discussing situations with your adult child. You and your partner should first agree on your position, and then stand firm together.

Keeping Them Under Control

Some parents may feel tempted to "keep control" over their adult child by controlling her finances. Renee, a 25-year-old medical student, had it made—or so she thought. Her parents bought her a $2 million condo in Newport Beach and paid for all of her living expenses—as long as she followed the life path they chose for her. Her parents were footing the bill, so Renee was a well-cared-for puppet on a string.

She went on to become a successful cardiologist, but she continued to seek her parents' approval for everything she did. She became completely dependent on Dad and Mom to tell her she was heading in the right direction. Sadly, Renee never explored other sides of herself, other passions that weren't on Dad and Mom's life agenda for her.

This family formed patterns that simply couldn't be broken. Renee's parents were anxious about giving up their little girl, fearing she might fail if left to her own devices. They wanted to control her—and they did so by financing her life. Renee learned to seek her parents' approval in every aspect of her life, even after she launched a successful career. And by not allowing herself to pursue other passions, she set herself up for a midlife crisis!

As we discussed in Chapter 5, sometimes parents will consciously or unconsciously hold a child back. If you find yourself influencing (or even controlling) your child using money, examine your motives. Take a close look at why you're holding on to your adult child. Are you afraid of being alone or losing your "job" as a parent? Do you want to protect your child from making a blunder with his finances or his life choices? Talk to a friend or your partner, or seek family counseling to help you sort through your feelings. The ability to be financially independent from you is one of the best gifts you can give your adult child.

The Least You Need to Know

◆ In an era of easy credit, Americans tended to spend more than save. But this trend is reversing as families work together to unwind debt, planning and saving for shared goals. Open nesting can be an opportunity to teach good money habits and put them to practice as a family.

◆ Young people today remain optimistic about the future, but many lack the skills to manage money successfully.

◆ Money is the number one cause of conflict between partners.

◆ If you're tempted to bail your child out, enable your child's bad habits, or control your child through money, remember that doing so could put your child's development and your future financial health at stake.

Chapter 15

You Look Too Young to Have Grandkids!

In This Chapter

- ◆ What grandparenthood looks like today
- ◆ Exploring grandparents' role in the open nest
- ◆ What does it mean to raise your grandchildren?
- ◆ Brainstorming easy things to do with your grandchild

Cultural anthropologist Margaret Mead wisely said, "In the presence of grandparent and grandchild, past and future merge in the present." While it's true that the connection between a grandparent and a grandchild is a truly beautiful and unique relationship, you probably didn't anticipate you'd be living together in your home!

A relatively small but growing percentage of open-nesting households include an adult child's child (your grandchild). It happens more often than you'd think. According to the U.S. Census' "America's Families and Living Arrangements: 2008," about 9 percent of all children—that's 6.6 million—lived in a household that included a grandparent. That means almost 1 out of every 10 children lives with a grandparent in the home.

Grandparents and grandkids under the same roof is an important issue to address. As the economy tightens, the number of single-parent households increases, and more women go to work full time, baby-boomer grandparents, or what the AARP calls "grandboomers," will play a critical role in caring for and even raising grandchildren.

In this chapter, we explore the grandparent/grandchild relationship and discuss what's appropriate and optimal when your adult child moves back with her own children in tow. We'll talk, too, about how to strengthen and nurture the relationship with your grandchild—and maybe spoil him a little, too!

Grandparenthood Today

Grandparenthood is a time in your life when you can enjoy and partici-pate in a young child's life, and from it gain a renewed sense of vigor and vitality. Whether you live with your grandchild or live thousands of miles away, the sight of a tiny smile or the sound of a teenager's laughter undoubtedly melts your heart.

To a grandchild, there's nothing quite like grandparents. Grandma will bake a batch of chocolate chip cookies just for little Sara. Grandpa will read Noah his favorite bedtime story four times in a row—even if it's an hour past his bedtime. When they get older, Grandma will teach Noah the names of every plant in the garden, while Grandpa will show Sara how to drive a pick-up truck when her feet reach the pedals. It's a very special and unique relationship.

Under Your Roof

The funny old saying "Grandparents and grandchildren get along so well because they have a common enemy" may be true, but it points to an issue that may surface: *triangulating* your grandchild. Watch out for focusing on your grandchild in order to avoid conflict with your adult child. And make sure never to criticize your adult child in front of your grandchild.

In today's world, grandparents have (happily!) become more involved in their grandchildren's lives than in years past. Grandpa and Grandma are healthier and better able to soothe a crying infant or keep up with a traipsing toddler.

Childcare experts say that grandparents are the "glue" holding things together for working parents today. Many children are cared for by several different people, often going to preschool or childcare centers in the morning, for instance, then to Grandma's house in the afternoon, and finally getting picked up in the evening by Dad or Mom. Grandparents certainly play a key role in caring for and raising young children—something you already know!

From a family therapist's point of view, grandparenthood is a time marked by certain life-cycle tasks. As you progress through "middle adulthood," becoming a grandparent triggers certain developmental milestones, including the following:

◆ You fulfill your wish to survive through your own progeny, which ultimately helps you accept mortality. Yes, it's kind of morbid, but it's part of your life cycle. By seeing the next generation sprout, you accept that your legacy will continue through them.

◆ As part of reviewing your life achievements, grandparenting enables you to relive your own child-rearing experiences from years ago and come to accept your achievements and come to terms with your failures.

◆ Being a grandparent also gives you the opportunity to reconnect and heal old relational wounds with *your* parents as you identify (and muddle through) challenges inherent in parenting. You begin to understand and empathize with what your parents went through with you.

◆ You also get the chance to bond with your grandchild in a way that's not complicated by responsibilities, obligations, and conflicts that naturally occur in the parent-child relationship.

Being a grandparent certainly has its ups and downs, but it's comforting to know that while you're sharing your life with him and her, you're also developing and evolving through your own life-cycle goals.

A Grandparents' Role in the Open Nest

So, you've opened your nest to your child and grandchild. What is your role in this scenario? Will you continue to be a doting Granddad, or might you need to take on more of a "serious" role as your grandchild's primary caregiver? Every household is different. Your child may need you to babysit your grandson twice a week when she's in class, or you may need to watch him every afternoon after daycare while she works.

In an open-nesting household with grandchildren, you should continue to nurture your adult child, and be a comfort, a friendly ear, and a helping hand—especially as she struggles to gain independence *and* raise a child. By helping your child with caregiving, you can play a vital role in helping her gain the confidence to re-launch with her child.

For your adult child, her primary relationships are with her partner and her child. She and her partner make up an executive system for their own family, and they will make the majority of decisions relating to the child—including how he is raised and disciplined. You're going to need to recognize and respect that.

Discipline and Feedback

When you raised your child, you had your own parenting style (and you raised a fine young man or woman!). Though you may be partial to your method, different parenting styles can yield good results, even if it's not the same approach you used. For instance, you may have swatted your son's behind when he misbehaved, but your daughter prefers to use "time-outs" in the corner.

New parents often raise their children in ways that are very different from their own childhood experiences. They have their own ideas. They have critiqued the parenting styles of their own parents and chosen which elements to keep and which to change.

No matter how tempted you may be to share your parenting wisdom, you should avoid telling your adult child how to do her job. It's tempting to pull rank and start giving directions again! Plus, you may so love the experience of having a small child or teenager in your home again that you run the risk of appropriating your child's parental duties under the guise of making things easier for your adult son or daughter. Be careful of overstepping your bounds. You can offer feedback, but remember that the final decision belongs to your adult child and his partner. Besides, they might not want your advice, anyway!

Under Your Roof

Though you should generally refrain from telling your adult child how to discipline her child, you should definitely intervene if the child's safety is at risk—for example, if the situation is one of potential neglect or physical harm to the child, or when frustrations cause tempers to flare. Grandparents can be welcome buffers and stress-reducers to relieve their kids from some of the relentless demands and pressures of parenting.

If you do completely disagree with something your child is doing, don't mention it in front of your grandchild. That can cause her discomfort, confusion, and to feel conflicting loyalties. The child wants and needs to love both of you, and not feel like she has to choose sides or please one person at the expense of the other. Instead, you and your adult child should present a "unified front" to your grandchild. Discuss child-related issues away from her, come to an agreement, and then allow your adult child to present the plan to his child.

Babysitting Duties

As we mentioned earlier, as many as 50 percent of grandparents in the United States provide some kind of childcare to their grandchildren. If your adult child and her son or daughter is back in the nest, it's likely that you'll be called upon for babysitting duty.

When you and your adult child formulate your open-nesting plan, which we discuss in Chapter 17, you should also create a plan or agreement outlining what you and your child's expectations are with regard to caring for your grandchild. It may seem silly, especially when it's "all in the family," but you'd be surprised by what things may need to be clarified. The agreement should include things like the following:

◆ **A schedule.** Though this will probably change while your adult child is living with you, try to come up with a schedule that works well for everyone.

◆ **Payment.** Obviously, this is negotiable. You probably won't charge to babysit your grandchild, but if you're foregoing income in order to help, it would be reasonable to discuss payment.

◆ **Emergency contact information.** This should include things like your child's cell phone number; medical information like the doctor's name and number, allergies, or medications; and who to call in case of emergency.

◆ **Nutritional guidelines.** If your adult child has her son on some dietary restrictions, like no sugar or junk food, know that in advance and follow the instructions.

◆ **Activity guidelines.** Your daughter may want her child to spend some time outside every day, be artistic, practice piano, or memorize vocabulary words. She may want TV or Internet time to be limited, or content to be monitored.

◆ **Discipline protocols.** If your son prefers "time-out" restrictions to swats on the behind or a raised voice in reprimand, he should include specifics like this in the agreement—and you should follow suit, even if you don't believe in it.

Revisit this plan as often as you need to be sure the childcare continues to go smoothly. You should also plan to have regular "meetings" with your adult child—like parent-teacher conferences—to discuss any concerns, progress, and behaviors at least once a month. And, of course, speak with your adult child daily about any conflicts, challenges, or disagreements that may have come up.

Ground Rules

If caring for your grandchild becomes too burdensome, talk to your adult child about it and come up with another solution. It's okay for you *not* to want to care for your grandchild 24/7! You still have your own life and interests, after all. Plus, you may not have as much energy for parenting as your son or daughter does. Draw the line if grandparenting duties are getting to be too much—and don't feel guilty about it.

When Grandparents Raise Grandchildren

In the United States, growing numbers of children are being raised solely by their grandparents. According to the 2000 U.S. Census, 2.4 million grandparents reported being responsible for the basic needs of grandchildren living with them. About one third of them are raising their grandchildren with no parent in the home. And 71 percent are under the age of 60.

Why are so many children being raised by their grandparents? The AARP's Grandparent Information Center found factors that play the biggest role include drug and alcohol abuse, mental illness, incarceration, death of a parent, poverty, divorce, and child abuse or neglect.

You may find yourself in a situation where your adult child is no longer able to raise her own children, be it permanently or temporarily. If that's the case, you *can* raise your grandchildren, but it will take some physical and emotional effort to do so.

Clarify Who's Who

Being raised by a grandparent can be tough on a child of any age, even if it's a temporary arrangement. It becomes ambiguous who the "primary" parent is. It used to be the adult child, but then it shifts, making the grandparent either a co-parent with the adult child or even the primary parent.

When rearing the grandchild, grandparents may try to follow the adult child's parenting model, but over time they inevitably will follow their own parental instincts. They may feel they need to "replace" the grandchild's biological parents in an emotional sense, but instead, they should try to remain grandparents, especially if the child's parent(s) expect to resume their roles at some point in the future.

The good news is that kids are incredibly resilient and can understand that one "parent" has one set of rules and another "parent" has something different. If you're raising your grandchild, be sure you communicate to him that you are not his *parent*, but you are his *grandparent*, and you have your own ways of doing things.

> **Ground Rules** _____
>
> One way to help keep relationships clear is to develop positive rituals that include your grandchild's parents. With luck, you'll be able to have phone calls or videos for your grandchild from a mom or dad away serving in the military or traveling for work. If direct communication is not always possible, then play games that incorporate family photos, or create a bedtime ritual that includes sending love and prayers.

Give the Child Time to Adjust

When a grandchild moves in with her grandparents, she'll need some time to get used to the new arrangement. She'll need to adjust to her parent being gone, living in a new place, and learning the new parenting style of her grandparents. Your son, for instance, may have allowed his daughter to eat dinner in front of the television, while you require her to eat at the table with the family.

Your grandchild may not understand why her parent isn't living with her any more. She may feel abandoned or rejected by her parent—or even wonder how much (or if) her parent even loves her. It's critical for you and your adult child to take every opportunity to remind the young person how much her parents love her and care for her. Explain to her that even though her parents are away fighting for our country, for instance, that they love her and want the best for her no matter what.

Grandparents, you should be prepared for the child testing her boundaries, too. She could act out or try to manipulate you. When this happens, be loving and understanding, but show the child that you have consistent limits and boundaries. Make the rules clear to her, and be sure she knows what the consequences will be for misbehavior. Only threaten consequences that you are prepared to follow through on. If your grandchild learns that you won't follow through (even sometimes), she will test her limits even more.

When faced with a difficult situation or big transition, some children could unintentionally regress by acting younger or losing acquired skills, like starting to wet the bed again. This is a tough situation, and you'll likely benefit from some family therapy, at least to get you through the initial hurdles.

Ground Rules

Encourage your grandchild to share her feelings about having to come live with her grandmother and grandfather through dialogue, art projects, play time, or writing. You should try to empathize with the child and resist becoming defensive and dismissing or "explaining away" the child's feelings. Try not to feel hurt when your grandchild misses her parents. Even with a "perfect" grandparent filling in, her parents can never be replaced.

Life with Your Grandchild

Keeping up with your grandchild can be a challenge. You're not as young as you once were, and all that activity could take a toll on you. To make sure you have the stamina to follow your little Energizer Bunny or active adolescent while keeping him engaged (and not tearing through the house), here are some ideas for making the most of your time together:

◆ **Hit the open road:** If your adult child is okay with excursions, take your grandchild on a destination adventure. Drive to the beach to collect sand dollars. Go to the forest and gather pinecones. Go fishing or bird-watching or camping. They're all relatively low-impact activities, and they'll keep his mind engaged—and his body busy.

◆ **Explore the past:** Share pictures and stories about your youth with your grandchild. Maybe you lived in a foreign country as a girl. Perhaps you moved several times a year, thanks to your father's job. Tell her about what you've seen and done with your life, and have her compare it to hers. You may find you have common interests, which can truly create a bond between you and your grandchild.

◆ **Cultivate the creative side:** What child *doesn't* like to scribble and paint and mold with clay? Or sing and perform? Encourage your grandchild's creativity (especially because kids don't get nearly enough of it in school!). Hang his pictures on a special "art gallery" wall. Help him make handmade cards, or compose and perform special songs for his parent's birthday or holidays.

Nurturing creativity in childhood can lead to an interest that might become a lifelong pursuit, or even morph into a career.

♦ **Put her to work:** Housework or gardening or pet care *can* be fun! Give your grandchild her own set of cleaning or gardening tools, turn on her favorite pop-music star, grab your own feather duster or trowel, and get to work. If you make it a fun and upbeat experience, she'll learn to enjoy working with her Nana to keep the house tidy and neat and the garden weed-free—and cut down on your own chores!

♦ **Play some games—video games:** Yes, really! Video games have come a long way since your Atari and Pac-Man days. Some are educational, some are sporty, and some are just plain fun. If your adult child or grandchild doesn't already have a Wii, Xbox, or PlayStation, invest in one and get gaming!

Under Your Roof

Be careful when it comes to play! Mature women can be prone to the brittle bones of osteoporosis or to conditions such as arthritis that make play a challenge. Men especially need to be particular about heavy lifting (you're not a young athlete whose youth alone is his greatest strength!) to protect from back or knee injuries. Certainly, get out there and experience the world with your grandchildren. Keep moving. Just be proactive and smart about protecting your health and avoiding injuries like falls and sprains.

The Least You Need to Know

♦ "Grandboomers," like you, will play a vital role in tomorrow's youth. Your adult child will likely call on you for duty.

♦ Be active grandparents.

♦ In an open-nesting household with grandchildren, it's critical for you—the grandparents—to adhere to your adult child's parenting style, even if you don't agree with it.

♦ If you find yourself temporarily or permanently raising your grandchild, give emotional support but set firm limits.

Chapter 16

Guess Who's Coming to Dinner ... and Staying?

In This Chapter

- ◆ Tackling tough issues like race and ethnicity, religion, politics, and sexuality

- ◆ Parental strategies for handling conflicts that come from hot-button issues

- ◆ Encouraging parents to be tolerant and understanding— even if their children aren't

- ◆ It's true: if it doesn't divide us, it unites us!

Families come in all shapes and sizes. As the family grows and evolves through time, differences of race, ethnicity, politics, religion, and sexual orientation, to name a few, can certainly challenge everyone to find new ways to grow together.

When young adults launch into the world and develop their independence, sometimes they'll explore things that are *different* from their family of origin. It's normal for them to do this, but

those differences could cause some disagreements in the nest—
especially when the adult children move back home with their new
discoveries.

As they enter into relationships of their own, they may also introduce
new—and controversial—points of view into the family's reality: a boyfriend from a different culture; a best friend with a different political
ideology; a girlfriend with a crush on your daughter. How do you as a
parent handle it? What's the best way to accept these differences?

In this chapter, we'll look for strategies to prove true that what unites
us is greater than what divides us. And if differences prove too great to
open the nest fully, we'll help you find ways to provide support, assistance, and love.

Forbidden Subjects?

Many a wise person has said to never talk about politics or religion if
you want to make friends. Mostly, they're onto something. Everyone
has his or her opinions about the "true" religion or the "correct" political point of view. We tiptoe around the issues of race, ethnicity, and
sexual orientation. Bringing up topics like these will almost certainly
spark a debate—if not a downright argument.

Remember *All in the Family*? The show's popularity surged because it
placed a glaring spotlight on some very controversial issues of the day—
politics, war, religion, women's rights, racism—issues that we're still
contending with today. Though as a society we're getting *better* about
accepting differences, we're not 100 percent there yet. We've elected
an African American president. We accept a variety of religions and
ethnicities. Some states have even allowed gay marriage. And politics?
Well, people will never agree about politics!

Generally, when dealing with issues surrounding race and ethnicity,
politics, religion, and sexuality, tolerance and acceptance are the keys
to understanding. Try not to judge your child or her friends about their
different opinions. It's okay to have varying points of view on controversial topics, but it's not okay to berate someone for her beliefs. You
are *family* after all, and by exploring your differences, you may find
yourself closer to one another in the end.

Racial Tolerance

The United States is a melting pot filled with people of different races and ethnicities. If you're like most Americans, you're probably a mix of several different ethnic groups. That blending of cultures is what makes our country so rich and diverse—but, ironically, we're not very tolerant of people or cultures that we perceive are different from us, or that we perceive have different goals and values than we believe we do.

If it's not important in your household already, ethnic and racial tolerance—rather, cultural acceptance—should be high on your list of priorities. By learning about different cultures, you enrich your own life and your family's individual "culture." You might be surprised to learn that we're not so different after all!

> **A New Generation**
>
> Webster's Dictionary defines *ethnicity* as a population subgroup having a common cultural heritage or nationality, as distinguished by customs, characteristics, language, and a common history. *Race* also refers to different people groups, but it groups them by physical traits, like skin color, hair, eyes, and body shape.

Race relations and acceptance in an open-nesting household may arise in a number of situations, which we outline here. You may not find yourself hotly debating the virtues of Polish immigrants or equal rights for African Americans, but you or someone in your family could unknowingly be stereotyping Middle Eastern culture and insulting your adult child's dinner guest. Here are some other situations that may come up:

◆ Your child may start dating someone from a different race or ethnic group, which could make you feel uncomfortable. As parents, ask yourselves why the person makes you feel this way. Try to genuinely get to know the person. Ask the friend tactful questions that show a genuine curiosity and respect (after talking to your child and doing a little research first, of course). Find things you have in common instead of focusing on things you *don't* have in common.

◆ As a family, you may find yourselves inadvertently making comments or stereotyping people based on the way they look or act.

You could be watching television and flip to the Country Music Channel—and start making jokes some might find inappropriate. You could be reading the newspaper and start making sweeping generalizations about those "immigrants who are taking away American jobs." Sometimes the associations might seem so innocuous that you could miss the idea that someone might be offended or hurt; for example, insensitive comments about adoption or even environmentalism could backfire potentially. Check yourself. Would you say those comments to the person if he were right in front of you? Remember the great dictum of medicine: *do no harm*. If the content of your comments is negative, think twice about saying those words that could hurt.

♦ Your adult child could begin to act prejudiced toward people of a particular race or ethnic group. If that should happen, talk to her like an adult. What triggered the sudden stereotyping? Does she understand how hurtful it can be? If possible, coach her through it and lead by example.

We are definitely getting *better* as a society, but try as we might, many of us still fall into the trap of prejudging people we don't know or understand. Rather than make those sweeping generalizations, we should focus on the *individual* and his or her character and strength.

Let's Talk Politics

Sure, we all know how our government *should* be run, right? Well, we certainly have our own opinions! We should have bailed out the banks; we shouldn't have bailed out the banks. We should get out of the Middle East; we should stay there and fight. Right, left, conservative, liberal, libertarian, green, our political viewpoints tell a lot about our beliefs, our ideologies, and what we value for ourselves and our country.

Which is right? Which is wrong? Who knows? And those loyalties shift widely depending on who's in power at a given time, and as the 1960s-influenced Baby Boomers, like yourself, get older, our country may lean one way or the other, depending on the dynamics of the current political climate. Regardless, politics is a hot-button issue and we have to remember that we *all* have a right to our own opinions—and that even includes your children!

A New Generation

According to a Rasmussen Reports poll in February 2009, 40.8 percent of Americans say they are Democrats and 33.6 percent say they are Republicans. Men are evenly divided between the parties—35 percent Democrat, 35 percent Republican, and 30 percent not affiliated; women are more widely divided, at 46 percent Democrat and 32 percent GOP.

Political debates in an open-nesting household can be healthy and mind-opening, or they could be painful and lead to raging fights that end with someone storming out of the room (or the house!). You probably know exactly what we're talking about. How many times have you sparred with your conservative friend about gun rights, or your liberal friend about reproductive rights?

If you and your family welcome those political debates in your household, here are some "rules to play by" to keep everyone in bounds:

- ◆ Make sure everyone gets a chance to express his viewpoints. Give each person an opportunity to speak—and try not to cut each other off.

- ◆ Rather than argue "right" or "wrong," listen to each person's opinion, and encourage her to back that opinion with reputable facts.

- ◆ Learn from one another. Junior may have read something on Drudge Report or Huffington Post that wasn't reported on CNN. You may learn something!

- ◆ When it gets too heated—or when flooding occurs, creating a temporary emotional overload—break up the debate. There's no reason to fly off the handle. It's only politics, after all!

In the end, it's okay to agree to disagree. Try not to let your emotions take hold and cause you to hold grudges against your child (or your parents). Conservative or liberal, we all want what's best for our country. We just have different ways of getting there!

Your God, My God, Our God

Did you raise your children with a certain belief system? Did everyone go to church, mosque, temple, or synagogue once a week, or did you teach your child values at home? Most families have some sort of moral guidelines that they use to raise their children, and many times they're centered on some kind of religion, whether it be Catholic, Christian, Buddhist, Jewish, Mormon, Islam, Wicca, or Paganism.

When adult children *differentiate* themselves from their family of origin and move on to be independent individuals, many times they abandon or change their religion. College graduates especially shift their spiritual preferences, and young adults may wish to experiment with different ways of expressing their faith. It's a normal part of development.

In your open-nesting household, how will the subject of religion be handled? If you attend regular services, will you expect your son to join you? If your daughter has changed religions or has married someone of a different faith, will you be tolerant? And if you have younger children still living at home, are you concerned about the example your adult child is setting?

When your adult child moves back home, you'll definitely want to broach these topics in an open and honest way in a family meeting, and you may even consider adding them to your child's open-nesting plan, which we discuss in Chapter 17. Here are some scenarios to consider:

- ◆ **Same religion, same devotion.** If you and your adult child still share the same religious beliefs and practices, great! Invite your son to attend services and proceed as you did before he moved out.

- ◆ **Dropped religion entirely.** If your adult child has banished religion from her life completely, you may feel tempted to bring her back into the fold. At this point, resist the urge. Show her love and compassion (and all the things your religion teaches you), but don't force her to do anything she doesn't want to do. When and if she's ready, she'll return. (Besides, it's between her and God anyway, right?)

  And what to do if your child is leaving provocative books all over your house, maybe a Christian existentialist classic such as

Kierkegaard's *Fear and Trembling?* Offer to read the book, and start a dialog about it.

♦ **Different beliefs, different venues.** Should your adult child change religions entirely, you may again feel like you need to "get him back on your side." This is a journey your child is taking on his own, and you should do your best to accept it and support him as he finds his own path.

Under Your Roof

Of course, if your adult child is involved in some sort of dangerous religious cult that is limiting his contact with you or draining his bank account, consult a licensed therapist and work together to remove your child safely from harmful influences.

If your adult child shows any kind of disrespect toward your religion, you have a right to say something. Use the relational approach that we discussed in Chapter 8 and tell your child how her behavior is affecting you. Remember that it's your house, and if you choose to pray before meals or study sacred scripture in your home, your child should show you respect.

Who Do You Love?

Civil rights for gay people is an explosive issue today. As our culture grows to accept homosexuality as a normal way of being right alongside heterosexuality, more and more individuals feel comfortable admitting they may be gay, lesbian, bisexual, or transgender. Look at the popularity of films like *Brokeback Mountain* and *Milk* or the success of television shows like *Will & Grace*. Being gay is going mainstream. Just look at the popularity of Ellen DeGeneres whose coming out raised controversy years ago, but who has grown to be beloved as the host of a popular talk show ... move over, Oprah!

Gay civil rights are even surfacing on the political scene. Gay rights groups across the country are lobbying to legalize gay marriage and give homosexual couples the same rights as heterosexual couples. Did you know that Ellen married her partner Portia de Rossi in August, 2008 in

Los Angeles? Though it's still a hotly debated topic, more American voters than ever are open to the idea of equal rights for all. (It wasn't long ago that women won the right to vote, after all!)

Older generations, however, may not feel comfortable with the whole "gay thing." They were raised with traditional values and beliefs—remember that "Leave It To Beaver" stereotype of husband, wife, and two children in a house with a white picket fence? That's what many parents hold true in their minds, and they don't understand that some people may be attracted to and fall in love with people of the same sex.

So, what do you do if your adult child comes out of the closet? Do you tolerate his gay friends? Do you allow her lesbian friends in your home? What if your daughter's best friend is a gay man? It can be confusing for parents, especially if you've never dealt with an issue like this before. It can be a process for most parents, in fact, to come to terms with their child's sexual preferences.

 Ground Rules

> Young men and women who have just discovered their homosexuality may feel like outsiders in their own home. Do what you can to help your adult child feel welcome. Encourage him to share his feelings with you. Accept her for herself, and not who you want her to be.

Here, we've listed some issues that may come up and how to handle them:

- **Coming out for the first time.** If your adult child suddenly announces that she's a lesbian, try to be as accepting as possible. It's very difficult for a child to share the news with her parents, and she'll need as much compassion and understanding as you can muster.

- **Trying to "change him back."** Resist the urge to try and "fix" your homosexual child. If he has come to terms with his sexuality, you should, too. Many young gay men, for instance, find it a relief to finally "come out," and if you keep trying to "put him back in," that could cause anxiety in his life.

◆ **Welcoming your child's friends.** Just as you'd show acceptance and tolerance to friends of other cultures, religions, or political viewpoints, do the same with your child's homosexual friends, too. They are *people*, and you can have a perfectly normal conversation with them that has nothing to do with their preferences for men or women!

In every case, it's critical for you to show compassion and understanding rather than judgment and condemnation. Even if you don't "believe" that it's okay to be gay or lesbian, it's not okay to push that ideology on to your adult child or his or her friends.

Accepting the Differences

If you're like most parents, you had a dream for how your child's life would unfold. You concocted a series of stories, or narratives, about how future events would play out. When your daughter was a child in pigtails dissecting a passion flower, you may have envisioned her as a world-class botanist. Instead, she moved in with her girlfriend and took a job at a local garden center. You dreamed your oldest son would marry a down-home girl and take over the family business. Instead, he took up photography, moved halfway across the country, and married a woman from Thailand. Those dreams you had for your child don't always pan out.

When you realize that your adult child has made a choice that differs from your dream, it can be difficult to accept. A very wise therapist once said that the crux of therapy is to give up hope that others will change. It's a process that can take many, many years. Your child has chosen her path and her partner, and neither is likely to change (especially on your behalf). Ultimately, it will be easier for you to accept those choices rather than resist them. With that acceptance will come a closer relationship with your adult child.

Some parents who can't accept their children's choices have a strained relationship with them—and some lose their relationships entirely. Your adult child should never be made to feel the guilt, sadness, and anger that accompanies your lack of acceptance of her. She should not

be made to feel that she has to choose between her family and her partner. Every person's purpose in life is to be true to themselves, not to
live by someone else's rules.

Just like with any major loss, you will need to take some time to heal
from the disappointment of your lost hopes and revise your dreams (or
create new ones). Talk to your partner, close friends, or a counselor who
can help you through this difficult process. Build a new narrative that
incorporates the new person or situation. What will your child's life
be like with this new partner? How will your life be different? Focus
on the positives and challenge the negatives. Does it matter what the
neighbors think?

An important part of accepting your child's partner and rewriting
your narrative is to open your heart and mind. *Really* get to know your
child's partner. Spend as much time with him as you can and express
genuine, non-judgmental interest in him and his life. Focus on the
qualities he possesses that you can appreciate and relate to. Make him
feel welcome in your presence. Your child will appreciate it!

We think that this simple advice may prove useful in many situations
where tolerance and acceptance are the goals: *do unto others as you would
have them do unto you.* Remember that you love each other, and all
things are possible with love. Should your differences prove too great
to keep you together under one roof, make the decision that you'll continue to look for ways to find common ground, to nurture healing, and
to be a loving family.

The Least You Need to Know

♦ Differences about religion, politics, sexuality, and race and ethnicity can be resolved in an open-nesting household, as long as everyone is tolerant and understanding.

♦ Prejudice and stereotyping still occur in our culture, but we're
  getting better about it. We should all try to celebrate our differences and look for our similarities.

♦ If your family can't agree on politics, that's okay! Political debates
  are fun—as long as everyone's individual points of view are
  respected.

Part **5**

Learning to Move On

Every bird must fly the nest, and though it may not seem possible to you at this moment, you *will* all go on to explore your separate paths. Opening the nest has given your family the opportunity to grow, perhaps through adversity. It has presented you with chances to learn about your family, and about yourselves at the same time. It has challenged you to work together toward shared goals. Now, the challenge is to let go and to embrace the future and all it holds—for you, and for your children.

Chapter 17

Making a Plan

In This Chapter

- ◆ Creating an open-nesting plan to fit your family's style
- ◆ Issues to consider when developing your plan
- ◆ Deciding when the "launch" will be
- ◆ Strategies for conducting a productive family meeting

By now, we know you've figured it out. *You need a plan.* An open-nesting plan helps put everyone on the same page by articulating each person's needs and expectations up front. It's much easier to outline those ideas first rather than do "damage control" once things veer off course!

Unlike a contract, a plan is looser and can be renegotiated as things go along and change. Most families will be successful with a verbally agreed-upon plan and will not need to go as far as writing up a contract. However, if you fear a plan won't be enough for your family (or if past experience points to this), you can refer to the more specific and more formal contracts in Appendixes A and B.

In this chapter, we'll give you some guidance as you decide what type of plan you need (detailed and structured or fluid and flexible), and craft the plan with your adult child. Hopefully, it will help ensure the success of your child's stay in your home—and one that helps ensure a successful launch back into the world. *You need a plan*, and we'll help you get started.

Look at Your Family

What's your family's style? Are your lives intertwined? Do you make decisions together as an organization? Are you involved in your son's day-to-day life? Do you consider your daughter your BFF? Perhaps the opposite is true. Do you stay connected with your adult child via the occasional e-mail or phone call? With family members living at home, do you scribble notes rather than actually *speak* to one another?

When it comes to creating an open-nesting plan, your family's style is important when deciding how detailed it should be. Each situation is different, so it's important for you to be honest about how your family functions and what your family needs. Many families will only need a rough outline; others will need a structured timeline with specific goals and achievements. Your family's open-nesting plan is your system or process for achieving an objective—launching your child back into the world. It differs from a contract in that the plan is fluid and flexible, while the contract states what you each agree to.

Remember back in Chapter 4, when we introduced Olson's Circumplex Model? As you'll recall, part of the model describes families' levels of cohesion and flexibility. *Cohesion* refers to how engaged the family members are with one another, and it's broken into four types: enmeshed, connected, separated, and disengaged. *Flexibility* refers to how well a family tolerates change and ambiguity, and it's broken into four types, too: rigid, structured, flexible, and chaotic.

The most ideal styles for open-nesting families tend to be *connected* or *separated*, and *structured* or *flexible*. This is a fancy way of saying that family members are involved in each other's lives, but not too involved. They're able to handle a little ambiguity and change, but they still are on top of what's going on. For these family styles, many times a simple, flexible plan is all that's needed.

Where open nesting can get a little dicey is with families that tend to be very involved in each other's lives or very distant from one another, and for those that tend to be too structured or too disjointed. If you see some of these qualities in your household—even just a little—craft the details of your plan with extra care and attention to each family member.

How "Official" Should Your Plan Be?

If your family tends to have a more rigid style, one person in the family may try to "lay down the law" and single-handedly dictate the plan. For instance, Dad might decree that Junior can stay three months—and no longer—only if he finds a job, saves $300 a week, and mows the lawn every Saturday. And if he fails at any of the tasks, he's grounded (or kicked to the curb).

Whoa! Encourage the rule-setter to take a step back. You'll want to work toward forming a plan based on an open discussion, complete with negotiation and sharing of feelings (in other words, by using a relational approach). A family with a more rigid style will need to fold some flexibility into the mix. These families may have some difficulty changing the plan from when the child last lived at home (e.g., in high school). The arrangement, however, will need to be different from when the child was a teenager. You'll need to take his new life-cycle goals into account—as well as your own. Let yourself off the hook in terms of having to be a very disciplining or active parent.

Families tending toward the chaotic end of the spectrum may need to make a special effort to sit down and make a plan, as planning and organization is typically not their forte. If Mom keeps her schedule committed to (a possibly foggy) memory, Dad's working hours vary every day, and Jane's classes and social schedule keep her running around, the family will need to agree to meet—and then *really* meet—to discuss how living together will work.

Look at guidelines for family meetings for help in structuring a discussion (which we list toward the end of the chapter). If your family tends to be a little chaotic, be sure to discuss how each family member's role will be different than last time the adult child lived at home, as well as what each member's expectations are for the living arrangements. Establish that Dad will cook dinner Monday, Wednesday, and Saturday and that Jane will clean the bathroom once a week. Doing this will decrease the anxiety that can be associated with a lack of clarity.

A New Generation

If you're in a household that constantly bickers, find comfort in knowing that a family that yells together can often laugh together, too. Sometimes, emotional intensity on one end of the spectrum means there can also be intensity on the other end. It can even be a good sign to be very emotionally passionate—it means that the range of emotions you feel and express is wide.

Planning for the Stay

Once you've figured out what type of family you are and how to approach forming a plan, you'll then need to figure out what issues to cover in your plan and how you should go about putting it together.

Issues to Consider

Throughout the book, we've introduced ideas to include in your family's open-nest plan. And here, we've summarized some of those issues for easy reference. They're by no means exhaustive; you likely have some items unique to your family and situation. But these offer a great starting-off point.

So when creating your family's open-nesting plan, ask yourself these questions:

◆ How will you determine the length of your child's stay? (A little later in the chapter, we talk more about establishing a timeline and/or accomplishing specific goals leading to the launch.)

◆ What "conditions" will you have (if any) in terms of what the child needs to be doing while staying at home? Does child need to have job? Apply to school? Save a certain amount? Stay clean/sober?

◆ How do you feel about your child's friends spending time at the house? Do you need advance notice? Would you like them to leave by a certain time? Will you allow boyfriends or girlfriends to stay the night?

◆ How will you divide the space in your home? Will your child have his own shelf in the refrigerator or pantry? Does he need to "reserve" the media room or computer?

- What personal boundary requests should you consider at home? Do you need to establish a "if the door is closed, don't disturb me" rule? If your child moves home with a partner, how will they (and you) ensure private space?

- Will you assign specific chores to your child? If so, what will they be?

- Will the family eat meals together or separately? Who will cook? Who will buy the groceries?

- How will you share expenses? Will you require your child to pay a percentage of the bills? If so, which ones, and how much?

- Are you going to charge your child rent?

- Will you financially support your child? What do you agree to? How long will you do it?

- How involved will you be in your child's life? Will you give your child the independence to make his own decisions (and mistakes)?

- Now that everyone is in a new life-cycle stage, how will your roles change? What do you each expect from yourselves and from each other?

- If your child moves home with her own child, how will discipline be handled? How much babysitting will you agree to? Who will be making the major decisions on parenting?

- Will you have regular sit-down meetings as a family unit? If so, how often? What day and time?

- In an age of Facebook and iPhones, how will you address privacy issues if your child is active in social networking and brings the Internet, literally, to your dinner table?

Again, feel free to address as many or as few of these topics as you need to create a functional, workable plan with your adult child. And remember that it can change as time progresses.

 Ground Rules

Make sure that you get input from your son or daughter and discuss the plan together. Everyone will know what's expected—which will make the open nest a more comfortable, livable place for everyone.

A Productive Process

The process of crafting a plan can be daunting. As you're putting your ideas together, try to keep these 10 elements in mind throughout the brainstorming process:

- **Be goal-oriented**. When crafting your plan, set specific goals (flip back to Chapter 3 for a refresher) with your child and come up with practical ways to achieve them.

- **Be realistic**. Don't set your goals too high. Examine the situation and be realistic about what everyone can achieve. And try to "catch" your family members doing something good; everyone loves praise.

- **Be positive**. Have fun with it! By keeping a positive approach, your conversations will remain healthy and productive—not reactive and defensive. Frame requests in terms of what you *do* want, not what you *don't* want.

- **Be specific**. Be as specific with your plan as possible (or as you need it to be). Some things, like whether you eat meals together, won't need to be carved in stone, while others, like paying rent, should be more planned out.

- **Be unified**. Everyone in the family should participate in discussing—and agreeing to—the plan. Iron out disagreements before they become bigger issues.

- **Be fair.** Make sure you're considering everyone involved, including siblings who are also living at home.

- **Be flexible**. Sometimes plans change. Be sure you can bend when need be.

- **Be aware of your family's style**. Is your family highly structured? Is it loosey-goosey? Know how your family functions and try to find a middle ground in terms of flexibility, as well as the cohesion, or level of involvement, between members.

- **Be creative.** Don't be afraid to "think outside the box." Hold your family meetings out on the deck with strawberry daiquiris. Let your child choose his own chores. Try different approaches.

- **Be prepared.** Expect that there will be challenges, and address them openly at family meetings.

Developing your family's plan with your partner and adult child will be a big undertaking. If it's too much to do in one sitting, take several days to rough it out. You may decide that each of you will brainstorm a chunk and discuss it together later. However you approach crafting your plan, remember that it will be worth it when it's complete.

Try to keep the process of planning fun and engaging for all family members: you want *everyone's* ideas and full participation. Once your children realize that the plan is one created *with* them and not imposed *upon* them, they will want to participate in its formation! Remember that even if the reasons for opening your nest had to do with a problem or difficult circumstance, the process of resolving that problem doesn't have to be stressful!

Planning for the Re-Launch

How long? That's probably *the* most important question to address when developing your family's open-nesting plan. How long does your adult child want to stay, and how long are you, the parents, willing to have your child under your roof?

For some families, establishing a time frame—three weeks, six months, one year—may seem like a logical way to approach the question. It's defined, it's absolute, and it's on the calendar. An adult child who thinks linearly or tends to "languish" at home will probably do well with a specified date for launch.

Other families may choose to approach the "how long" question in terms of achieving a goal rather than setting a specific time frame. For instance, your daughter can live with you until she has a down payment saved for her new condo, or your son can stay until he emotionally heals from a divorce. Most adult children will know what they need to do, and when they're done, they'll head out and move on with life. Micromanagement or "over-helping" from you should not be necessary for most adult children.

In some cases—probably a growing number as our population ages—adult children may return home to assist their parents through a health problem or a house-related project. If you have back surgery, for instance, you may need your son's assistance for a few months. Or if you're building an addition, you may ask your contractor daughter to move in for a month to help with the framing and drywall.

Under Your Roof

Though you want family meetings to be cordial, some may spark heated debates that could cause family members to start flooding, which is when the debate causes a temporary emotional overload. If that happens, pause the meeting for a half hour and resume it after emotions have calmed down enough to continue.

In cases like these, because the parents asked the adult child to move home, she should be sure to articulate her own boundaries. It's a different type of situation, but it requires a "plan," too—one that mainly involves how long your adult child is willing to stay. Your family can still make use of the approaches we discussed earlier in the chapter.

Family Meetings 101

Your family meetings, which we suggest throughout the book, are a time to share joys and frustrations with one another, bring up issues, solve problems, and keep lines of communication open. So how do you conduct a successful get-together? Try the following techniques. They'll probably remind you of those meetings in the conference room at work—but these tips will help them feel much more efficient!

1. Set aside an hour at a time that is convenient for all family members who are living at home. You may not need a full hour, but it's better to budget too much time than too little—and you'll soon learn how much time your family will *really* need. If a key decision-maker can't be present, then reschedule.

2. Determine how decisions will be made in the family. It is important to understand the family's decision-making process *before* meetings begin. Will it be a democratic process (majority rules)? Does one person tend to "take control," like Mom or the adult child? Do Dad and Mom discuss the issue as an "executive system" and come back later to give their "verdict?" If you don't know *how* a decision is going to be made, then it's much harder to resolve an issue in the meeting. Remember, now that your child is an adult, he should have primary decision-making power over issues that affect him (like job pursuits or what grad school he applies to). Issues that affect the entire family, like who gets to use the media

room on Thursday nights, should ideally be decided by consensus of all family members.

3. Decide on goals for the meeting. Set a realistic scope. Don't try to accomplish too many goals in one meeting. Make sure each person has an opportunity to come up with a goal—everyone, including younger siblings, needs to feel included. Some examples of goals include your adult son wanting to find a way to have more privacy when he's entertaining friends, or perhaps you want to solve the problem that food you plan to cook with keeps "disappearing" out of the fridge.

4. At the beginning of your family meeting before you launch into discussion, set an agenda. Family members should come prepared with discussion points to place on the agenda, and the designated note-taker can write down what topics will be addressed in the meeting. Make a "schedule" for the meeting, dividing the meeting time between the topics, and budget at least 10 minutes at the end of the meeting for "wrap-up" and summarization of the decisions.

5. Choose one person to start the meeting by presenting his issue to the rest of the family and stating what he is requesting.

6. Stick to the topic at hand. Thoroughly discuss each issue before going to the next, using communication techniques we introduced in Chapter 11.

7. While discussing a topic, don't spend the whole time talking about the problem. Make sure you allow enough time to talk about realistic solutions. Take a few moments to formulate a plan for how that issue will be addressed. Identify what the "next steps" should be.

8. Identify specific things each person will do to solve the issue. To make those action items a reality, refer to Chapter 3's goal-setting section for help. Talk about specific actions that people will do to solve the issue. They may need to be broken down into steps or be made more specific. Rather than speak in theoretical or general terms, outline realistic, practical steps to achieve the goals you lay out.

9. When time is up for one issue, move on to the next, with a promise to come back to the initial issue later in the meeting or to prioritize

it on the next meeting's agenda. Don't get bogged down. Make sure you stick to your agenda, and if you run out of time for one issue, too bad. Unless all family members agree to continue on the current topic, it's important to move on and address the other topics on the agenda.

10. At the end of the meeting, conclude with a summary of what issues were discussed, and ask each person to summarize what she plans to do to help. For instance, with the refrigerator problem, Mom might agree to label the food that she needs saved, and Jason might agree to not touch that food, or if he does, to replace it that same day with identical food. If follow-through seems to be a problem, or if family members tend to "remember things differently," write down what was agreed upon and keep the document in a file folder with similar lists.

Depending on your family's situation, try to hold a family meeting once a week or once every two weeks, with the minimum being once a month (unless your family is truly unusual in its level of harmony and organizational ability). Families in times of crisis could have several meetings per week. Try to schedule the meeting at a regular time— say, Sundays at 5:00 P.M.—so that family members can plan around it. It's okay for meetings to be rescheduled (within reason), but doing so should be more of a rarity than a regularity.

A New Generation

Don Tapscott, author of *Wikinomics: How Mass Collaboration Changes Everything* (Portfolio, 2008), asserts that now "it's not what you know that counts anymore; it's what you can learn." Many young adults today must quickly adapt their skills to constantly changing job requirements, like having to learn a new computer program, navigate through the newest social networking site, or use the latest piece of computer hardware. Lucky you! If you're not as tech-savvy as you'd like to be, your adult child might just be a sage resource for you.

Techniques like the following keep the mood of your family meetings light and allow for arguing parties to calm down, regroup, and approach the subject again:

♦ Using humor—a joke, a self-deprecating comment, or something that will make everyone smile

♦ Making a caring remark, like "I understand this is hard for you"

♦ Establishing common ground, saying, "We're in this together"

♦ Showing appreciation for the other person's feelings along the way, using visual or verbal cues to acknowledge understanding

By using a positive approach toward problem-solving family members can begin to feel …

♦ Safe in sharing openly with the group.

♦ Empowered to make confident choices rather than helpless.

♦ Validated and empowered rather than discouraged or negative.

♦ Listened to and accepted by members of the group.

♦ Focused on solutions rather than the problems.

Family meetings can be incredibly productive—or totally out of control—depending on how they're run. (You've probably sat through some chaotic, unproductive meetings in your lifetime!) By using these suggestions and personalizing them to fit your family's style, you'll be on your way to tackling issues before they turn into conflicts.

The Least You Need to Know

♦ Before you begin to formulate your family's open-nesting plan, determine what type of family you are. Is your style more rigid or more flexible? Do you tend to be more connected or more disengaged? Knowing your own bias will help you move toward the middle and create a plan that's right for your family.

◆ Some adult children will need to form highly structured and detailed plans, while others won't need much at all. Craft your plan based on your family's unique needs.

◆ Try to avoid limiting your child's "launch date" to a specific time frame. You can also use an achieved goal as the "time" to launch.

◆ Well-run family meetings will do wonders for connecting and communicating. Use them as an opportunity to work through issues—but also as a time to come together as a family.

Chapter 18

When They Keep Coming Back

In This Chapter

- ◆ Defining "adulthood" in your family
- ◆ Deciding whether it's reasonable for your adult child to move back home—again and again
- ◆ Why they may need to come home more than once
- ◆ Offering—and asking for—emotional support

While most of the time an adult child's sojourn back to the family homestead is a one-time event, sometimes a child will want (or need) to come home time and time again. For whatever reason, the young adult flies the nest—and boomerangs right back, just when you started enjoying your den or sewing room again!

Do you go along with it? When your son or daughter comes knocking over and over again, should you answer the door? Once the child has reached "adulthood," what is reasonable to expect in the parent-child relationship? You may feel obligated to welcome your child back the first time (or times), but when do you put up your "no vacancy" sign?

Becoming an "adult" and achieving true independence differs from person to person. Some children launch on their own with no troubles at all. Others need to make the transition in smaller steps, and still others need to keep Dad and Mom's as a home base—and keep returning when things get to be too much to handle.

Should you always have them back? Maybe, and maybe not. In this chapter, we'll first explore what being an "adult" really means in today's world. Then we'll help you decide what's reasonable, and explore some key issues that open-nesting households may face when the adult child keeps coming back.

An Age or a State of Mind?

What does "adulthood" mean? Is it when you reach age 18? Live independently? Own a home? Get married? Have children? Find a stable career path? Can someone be considered an adult *without* doing these things? Certainly, "adulthood" means different things to different people.

You could look at becoming an "adult" as reaching a certain age, or achieving physical or spiritual maturity. Different cultures and religions view these ages or levels of maturity differently, but many of them have rituals or celebrations, like a Quinceañera or a Bar Mitzvah, to mark the transition from childhood to adulthood.

In the United States, because our population is a melting pot of cultures and religions, we have no single "coming of age" ritual or milestone that demarcates adulthood. Instead, we tend to focus on chronological age. Young Americans are technically considered "adults" at age 18 (with full adult privileges coming at 21). When you turn that magic number, you're legal—and expected to obey the rules of the land and become a fully independent member of society.

You could also consider the concept of being an "adult" as accomplishing achievements, like getting married, having children, or establishing a successful career. Rather than reaching a magical age, a young person becomes an adult when she makes partner with her accounting firm or when he has his first child. Thanks to more young people pursuing graduate degrees, launching careers, and waiting for marriage, achieving those milestones is happening later for young adults today.

Ground Rules

If your adult child needs to move back home and into your open nest, this doesn't make her *less* of an adult. Try to keep an open mind. She may be making a very mature, well-thought-out decision to live at home for a while—especially in these times of higher unemployment and lower wages—even if it *is* the second or third time.

Suffice to say that the transition from adolescent to adulthood has prolonged, and many young people change jobs every other year or more. But do these trends mean that "adolescence" is extending, or simply that "adulthood" is defined differently, and is characterized by more exploration of relationships and careers?

What's Reasonable?

If your adult child keeps coming back to the nest, how do you determine what's reasonable? You can decide whether your child is achieving the developmental milestones that are appropriate for someone that age, which we discussed in Chapter 3. While your adult child is living with you, she should be making continued progress with these tasks—and getting closer to being independent.

Overall, if an adult child is indeed "making progress" in life, you shouldn't be too concerned if he comes home from time to time. Perhaps at one point he's between college and his Peace Corps assignment. The next time, he's between the Peace Corps and applying to grad school. The third time, he's waiting to save up a down payment for his own place. It doesn't necessarily mean he's stagnating or that he's having "issues" because he's moving home. He has a path and these are transitions in the middle of the path.

If an adult child is *not* "making progress" as you see it, that's a different story. Does she seem to be drifting along in life, perhaps overly depending on you to pick her up and dust her off? Does she seem unable to stay put in one career, or is she burning through money faster than she can make it or than you can give it to her … over and over again? Try to identify the reasons behind her lack of progress. Check out the next sections, where we've outlined some of the most common problems and suggested some solutions.

Under Your Roof

Don't expect that you'll be able to help your child through all his problems on your own. Sometimes, counseling can be a lifesaver. When a child is coming home over and over and is not "making progress" in his life, you can bet that something bigger is probably going on that you and your partner can't really address by yourselves. Anxiety, depression, addictions, or interpersonal problems that keep him from holding down a relationship or a roommate—these are things that call for help from a professional.

Times of Transition

Young adults today make myriad transitions, and many simply need a home base. As we discussed earlier, most young people hold a number of different jobs through their early adulthood, attend school and graduate school, and cycle through several "serious" relationships. When things change—when they're in between jobs, in between semesters, or in between partners—they come home.

When it comes to these transition periods, some issues may surface in an open-nesting household, from less-than-smooth moves to being caught on a seemingly aimless path.

A Bumpy Ride

A young adult can handle transitions in all different ways. He can go seamlessly from one job to another, or from school to work. He plans ahead for the "what-ifs" in life. He saves some money and comes up with contingency plans should he lose his job or break up with his partner. Some young adults may just be better able than others to plan for their transitioning interludes.

Others, however, may take some time in between ventures to "regroup." For instance, Nick, a 25-year-old pharmaceutical rep, faced a bumpy transition. As many young people do, he often only thought of what was right in front of him—work and socializing—and didn't plan for rainy days. When he was laid off, he didn't have any money saved so he had no choice. He had to move back home with Dad and Mom—but this wasn't the first time. It was his third. He moved home after college. He moved home when his live-in girlfriend dumped him. And he moved home when he lost his job. His parents allowed it, but they weren't happy.

If you're like Nick's parents and prefer that your adult child "seamlessly" make transitions between jobs/school/relationships instead of coming home in between, make sure he has the skills necessary to accomplish this. Start by making sure your child knows you expect him to think ahead and navigate a transition by figuring out what resources are needed to meet it. See if he has the budgeting and planning skills to handle day-to-day life, as well as to anticipate rocky times. If not, use your life experience to talk with him and help him plan.

I'm Back—Again

What if these transitions feel like déjà vu? What if there's a pattern of frequent job losses or break-ups, punctuated by stints back in your open nest? These days, living together before marriage is very common, so if your child splits from her partner, that could be a transition that brings her to your door.

Many break-ups and job losses are unforeseeable. Ideally, your child should have some savings to get by for a while if she needed to move out of an ex-boyfriend's apartment or search for a new job. However, the reality is that many young people live paycheck to paycheck. So the most important question to keep in mind is, "What does my child need right now?" Does she need to learn independence and personal responsibility (and that you won't bail her out)? Does she need emotional and/or financial support?

You could follow at least three possible paths when your adult child repeatedly moves back home:

- ◆ Let your child "learn the hard way" by not contributing money or opening your home. Maybe the child needs to learn personal responsibility or independence at this point in her life—especially if the same situation has happened several times already.

- ◆ You could subsidize your child for a certain period of time, enabling her to get back on her feet. Maybe she just needs to borrow $5,000 for rent and living expenses for a few months. If she's borrowed money before and repaid it in the past, she most likely will again.

If he's borrowed money and struggled to pay it back in the past, or not repaid it at all, then think about whether you can prudently prepay a service or fee, such as rent or tuition, directly to the provider on your child's behalf. Or try to find some other way than financing the endeavor, to help your child.

♦ Allow her to move in with you until she can regroup and figure out what she's doing next. If you choose this path, be sure to make a plan that outlines how she'll achieve her independence again— this time, permanently! (Flip to Chapter 19 for some tips.)

If your adult child finds herself transitioning often into your home, there's no single *right* way for parents to handle it. Sort out what your child truly needs and address that need. Hold in mind what life-cycle tasks are appropriate for your daughter at this time. For instance, maybe "learning the hard way" is what your daughter needs. She's 28, and she should be figuring these things out for herself at this stage.

A New Generation

Statistics from the College Board reported in *The Wall Street Journal*, reveal that families borrowed $85 billion in school loans for the 2007–2008 school year, compared to $41 billion 10 years earlier. Traditionally, and as recently as 2007, only about 5 percent of students defaulted on those loans two years after graduation. But those numbers are believed to be on the rise, with fears that they may hit the 30 percent mark last seen in the 1980s recession. Lenders are under pressure to renegotiate the terms of school loans.

Lost Path

When an adult child makes many disjointed transitions, he may appear to be aimlessly wandering through life and not coming any closer to finding his "path." He may see friends (and partners) come and go. He may try different professions. He may choose to travel or explore the world, coming home in between intercontinental jaunts. Many young people today aren't sure of what to do next, so they search—and feel comfortable using Mom and Dad's house as home base.

This can be disconcerting to parents, who inevitably want to see their adult child choose a path and excel at it. No doubt you want your son to settle down, start a family, and decide on a career. If you worry that your child fits the description of "aimless," consider these possible reasons *why* your child wanders:

◆ Is he simply an "explorer" who needs to try things out before seeing if they suit him? If so, encourage the exploration, but try not to enable aimlessness with excessive financial support or repeated open nesting. Teach him to plan for his transitions—keep a studio apartment, even if he's not living in it; deposit money in a savings account, even if he'd rather spend it.

◆ Does your child have low self-esteem and not believe she can succeed at a given path? Is she anxious or depressed, feeling dissatisfied with every path she tries? For either of these issues, professional counseling can be a lifesaver.

Transitions can be tough on everyone—especially if your adult child uses your home as his headquarters. To help your child ease through these changes, make sure you teach him life skills he'll need to succeed, and avoid enabling his sojourning tendencies. You'll all be better for it!

Trouble Planting Roots

Another potential problem that surfaces when an adult child moves home more than once is that she has trouble holding onto her own place to live—which results in her moving back home with you. She may lack the age-appropriate life skills to live on her own. This often happens for one of two reasons:

◆ **Financial trouble.** Your child may get laid off or fired from her job, or she's mismanaged her finances.

◆ **Difficulty getting along with roommates.** She may have interpersonal problems, like choosing unreliable friends. The child herself may be unreliable, not paying rent, not pitching in on the household chores, or not treating roommates with respect.

If your child has money troubles, counsel her about financial management. Recommend she read a money-management book (we list a couple in Appendix C), or have her meet with your financial advisor. Remember that one of her life-stage goals is to develop financial literacy and become financially independent, so this can be an opportunity to help your child problem-solve and develop skills she'll use in the future.

> ### Under Your Roof
>
> Drug use or abuse could cause a young adult to be kicked out of his apartment—either because of associated financial trouble, or the sheer illegality of what he's doing. If you feel your child has a problem with drugs or alcohol (or any type of addiction, for that matter), recommend—and pay for—a counselor skilled in helping people work through addictions. Most people have difficulty kicking their addictions without either professional help or a 12-step program.

Emotional Support

During times of crisis, be it health, emotional, or otherwise, families tend to come together for support. People feel safe and secure with their loved ones. Being with supportive family members helps a person work through crisis without fear of being judged or ridiculed. This is true for your adult child as well as for you.

Your Child Needs You

Sometimes, an adult child simply has an emotional need to come back to the comforts of home. Your son may need the support of his family after a relationship disaster or a devastating job loss. He may feel isolated, depressed, or anxious, and it's completely normal for him to sometimes reach out to family members for emotional support. If your child is doing this, you can consider yourself lucky to have a trusting and open relationship with your child!

However, if you worry that your child is relying on you excessively and is not able to handle being out on her own, help her find and pay for counseling. This is especially crucial if you are concerned that your child could suffer from depression, anxiety, low self-esteem, drug or alcohol abuse, or chronic relationship problems. (For more about depression, anxiety, and self-esteem issues, see Chapter 5.)

We'd be remiss if we didn't point out that sometimes our children's problems are mirrors of the challenges we've faced, or are facing, ourselves. There's that classic quote from psychologist Carl Jung: "If there is anything that we wish to change in the child, we should first examine it and see whether it is not something that could better be changed in ourselves." Family counseling can help to hold up this mirror and guide the family toward new behaviors and ways of being that are more supportive of life goals and present better paradigms for success—both at home, and out in the world. Extend your compassion toward yourselves as well as your children. Use patience and persistence to heal your child's emotional wounds, and perhaps your own as well.

You Need Your Child

So, what if you may need your adult child to come home to support you through a crisis? If you are suffering through a health issue, for instance, you may request that your daughter stay with you for a while. It's okay to ask her for help, but you should keep in mind that your child is in a life stage normally characterized by increased independence, so make sure you let her know that it's okay for her to leave when the crisis has passed, or when you've found other means of support.

Consider Emily, a young woman headed to New York City to begin a career in television journalism. When her dad was diagnosed with cancer, the temptation to call Emily home (and for Emily to feel that she should return to the nest) became very strong. Emily's mom's best friend took Emily aside on a visit home and told her to resist the pull—letting Emily know that her accomplishments and progress would ultimately bring more pleasure to her parents than the short-term relief and comfort of having their daughter back in the house at the cost of delaying her dreams.

Or, you may also feel you "need" your child to ease marital tensions you're having with your partner. If you and your partner discover your marriage may be in trouble, you might feel tempted to turn to your adult child and ask him to move back home, act as mediator or point of focus, and transform the household into a more harmonious place. As we discussed in Chapter 2, this is called triangulating your child, and it's not healthy for anyone. To ease marital tension, do not rely on your child. Visit a couples' counselor instead.

Another reason you may invite your child home is for help with household projects. You ask your child to help with a remodel, build a water feature, or add that tennis court you've always dreamed of—and you're willing to pay her top dollar because she was just laid off. In a case like this, it's best to hire someone else, even if you want to help your daughter financially. It's unhealthy to encourage overdependence on you. Instead, help your child find ways to find other employment and remain independent. It's best for her in this life stage.

One More Time

We'll leave you with another quote from Carl Jung: "Nobody, as long as he moves among the chaotic currents of life, is without trouble." Just like you will need to plan for an extended romp through the Golden Years (you're going to live to be 100, right?), your child lives in a society increasingly geared toward a prolonged experience of young adulthood.

You'll know when it makes sense to open your nest again … and again to your adult son or daughter. Be flexible and open-minded; remember that each trip back to the nest need not (and probably should not) be the same trip. Rethink the rules to fit each situation, and remember that oftentimes all your family may need is the grace of a second (or third) chance and the love to give or receive it.

The Least You Need to Know

- You can define "adulthood" by age or by achievements. In American culture, we typically deem someone adult when he turns 18, with full adult privileges at age 21.

- If your adult child wants to move back home over and over again, only you can decide what's reasonable and what's unacceptable.

- Young adults go through many transitions in today's world, and sometimes they simply need to consider Dad and Mom's house their home base—physically and emotionally.

- *Should* your child keep coming back? Examine the reasons why and help her sort them out. Help her develop the tools and skills she needs to fly from the nest—this time, for good.

Chapter 19

Planning an Escape ... Yours

In This Chapter

- ◆ Issues that may surface when you're trying to accomplish your life tasks and open nest
- ◆ Using goal-setting skills to stay on track
- ◆ Looking forward to the launch—for good!

You had plans and dreams and goals for your life ... your life *after* the kids were raised. And you were on your way to making those plans and dreams reality. Whether you were going to volunteer, start your own business, curl up on the sofa with your partner and all the good books you'd wanted to read all those years, or take a trip around the world, you were ready to move on with life when your children flew the nest.

Now that your adult child has returned, you may think you need to put those dream on hold. Not so! In fact, now is the time to embrace those goals, make those plans, and pursue those dreams so they *do* become a reality. Your life-cycle tasks at this stage of your life involve looking beyond your family and getting involved in your community. It's normal—and healthy.

This chapter explores ways to hold on to your own dreams when your adult child moves back home. After all, your child is not the only one who has an entire lifetime ahead! Being a good, responsible parent doesn't have to mean giving up your life (or your future) for your adult children.

Issues That May Surface and How to Solve Them

It's not easy focusing on your own goals and life-cycle tasks when an adult child moves back home. You'll probably fall back into old habits and routines. You'll become Mom or Dad again, taking on those old roles like cooking the family meals or disciplining your child like she was in high school again. Issues will surface, and we've listed some common ones here, along with some ways to handle them.

Heading in the Wrong Direction

When your adult child moves back home, you may be tempted to focus your energy on him, rather than moving forward with your own dreams. Thomas and Sandy, for instance, found new freedom when their son, Matt, went off to college. Because they had saved through-out their careers and were able to purchase several rental properties, Thomas and Sandy were able to "retire" early and follow their dream of spending a year driving their top-of-the-line R.V. across the country. Thomas mapped the route and the timetable; Sandy packed the rig and stowed the supplies, and pulled out her old 35-milimeter to photograph the journey. They found short-term renters to move into their home while they were traveling. They were ready to go!

Then the phone rang. Matt needed to come home. He didn't get along with his roommates, he was failing his classes, and he was running out of money. Matt wasn't prepared to survive on his own quite yet. So Thomas and Sandy put their trip on hold. Their son needed them, and they felt they had little choice but to turn their focus from their own aspirations to their son's troubles. Though they loved their child, Thomas and Sandy felt a tinge of resentment that slowly grew once Matt moved back home.

In this situation, Thomas and Sandy were ready—and eager—to move on with their lives. Thomas wanted to enjoy his leisure time and travel across the country. Sandy wanted to photograph the country and pursue her lifelong dream of publishing a travel book. But when Matt needed help, they felt obligated to give him what he needed.

> **A New Generation**
>
> According to statistics from the AARP's Multi-Generational Housing Patterns survey published February 2009, 15 percent of respondents live with their children more than 18 years of age.

Rather than drop everything and rush to Matt's aid, Thomas and Sandy should have continued with their plans. They could have helped Matt problem-solve through his issues and resolve them on his own while they were traveling across the country. They could have coached him through talking with his roommates and solving any interpersonal conflicts. They could have recommended Matt take fewer classes, or find a part-time job.

Just because your child needs your help doesn't mean you have to put your life on hold. There are ways to help from afar! Of course, if your child is truly in need, by all means, help him. But don't veer too far off course, whether you're traveling across the country or planning to submerse yourself in a hobby.

Trouble "Getting into Character"

Parents may find it difficult to shed that "Dad" or "Mom" character and treat their adult child like an adult. Heather, 27, moved back in with her parents, Dan and Janice, for a short while after returning from serving in the Peace Corps. She was saving money for a down payment on a condo and figured she would only need to stay six months at home. Dan and Janice welcomed her with open arms. They loved their "parent" roles—Janice, a stay-at-home Mom, had home-schooled Heather and her brother through high school, while Dan lived to provide his wife and children with everything they needed to thrive.

When Heather returned home, Dan and Janice fell right back into their old "jobs." Dan resumed "protective dad" mode, keeping track of where Heather was and whom she was with. Janice, who enjoyed regular meetings and lunches with her friends since the kids had launched, put those on the back burner, instead staying home 24/7 to be there for her daughter.

Heather, however, didn't need—or want—her parents to coddle her and treat her like a child again. She had just survived two years on her own in Cameroon teaching English to orphaned children, so she had developed her independence (in a big way!). What Heather needed was some *interdependence* with her parents. She wanted to move the relationship from parent-child to adult-adult. Interdependence, which we discussed in Chapter 3, is when parents and the adult child remain independent yet able to rely on one another for mutual assistance or support. It's an ideal relationship—especially in the open nest.

In this case, the parents had some work to do. When Heather launched the first time, Dan and Janice were well on their way to embracing the next phase of their lives. Dan was backing off his work schedule and enjoying less parenting responsibility. Janice was deepening relationships with her friends. But when their daughter returned, the old roles did, too. Heather was ready to treat her parents like peers, but they weren't ready (or able) to reciprocate.

Under Your Roof

During midlife, a disconnect can occur between partners. As women tend to develop more autonomy and get increasingly involved in their communities, men tend to become more focused on their relationships—and less on their achievements—as they age. Men may want more time from their wives to travel or participate in leisure activities—time the wife no longer has! If this happens in your marriage, talk to each other about your expectations, plans, and dreams. Try to negotiate and work out a compromise so you both have your needs fulfilled.

Old parenting roles are easy to fall into, especially if you really loved being a parent! If you find you have trouble "getting into character"—that is, a *new* character—when your adult child moves back home, remember your life-cycle tasks at this point in your life. Evaluate your family dynamics and try to give your child space to be independent—and give yourself room to follow your dreams.

Consumed by Guilt

Parents may also feel guilty about not giving their adult child the attention he needed early on, so they may overcompensate by becoming overly involved in his life. Grant and Barbara both had full-time careers while they raised their son, Christopher. Growing up, he was pretty much on his own—he got himself up and dressed every morning, he made his own breakfast and lunch, he rode the train to school, got himself to football practice, and found his way home. Before his parents returned from work, he was expected to have the table set and dinner warmed up. They ate together, but then Christopher would head off to his room to finish his homework. Needless to say, their family was disengaged.

After Christopher moved out on his own, Grant and Barbara both retired, and realized how little time they had spent with their only child. They were so focused on making a living that they didn't take time for the life they had with their son. So when Christopher needed to move back in after his girlfriend left him, Grant and Barbara saw it as a second chance to live with and enjoy their son. They took their disengaged family dynamic of the past and reversed it—they became overly involved. They wanted to know everything their son was doing, where he was going, and how they could be involved in his day-to-day life. Being as independent as he was, Christopher high-tailed it out of his parents' house as soon as he could.

These parents were consumed by guilt. They felt so bad that they missed out on their son's life, they overcompensated—and drove him away! Grant and Barbara raised a highly independent son, but they didn't develop the interdependent relationship that would have allowed the open-nesting situation to be successful.

If this sounds familiar, resist the urge to hold onto your child too tightly. The past is the past, and all you truly have is *right now*. So what can you do right now for your adult child? Encourage him. Treat her as an adult. Seek that interdependent relationship that develops from trust and confidence in one another. Don't let guilt ruin the relationship you already have with your adult child.

A New Generation

The Baby Boomer generation represents a big chunk of the American economy, and marketing gurus—from pharmaceutical companies and insurance programs to clothing designers and beauty product makers—tailor a lot of their messages to you. Though they traditionally temper their message to those over 50 years old, your generation—and its more than $2 trillion in assets—has a lot of buying power (which we hope you still have!).

Staying on Track

Staying focused on yourself and your goals when your adult child moves back home with you will take discipline. Your relationship with your child is likely a positive one, so it's natural for you to want to resume your parent role. It's important to remember, however, that as your adult child develops her own independence (even if she is at home), you'll need to back off, relinquish that parental control, and move on with your life.

To do that in a way that will keep you on track, consider setting specific goals for yourself, and for you and your partner, and for your family as a whole. In Chapter 3, we discussed how your adult child could set goals to achieve her independence. You can use the same goal-setting techniques to help you stay true to your life-cycle tasks rather than setting them aside when your child is back in the nest:

1. Determine what life-cycle task you want to focus on. Perhaps you want to get more involved in your community, or you want to deepen relationships with your friends.

2. Define that goal more specifically. For instance, you would say something like, "I will volunteer once a week at the farmer's market," or "I will play golf once a month with my friends."

3. Break the goal into steps or benchmarks. This helps to keep you on track toward achieving your goal. If your goal is to volunteer at the local hospital, for example, you would need to first contact the volunteer coordinator, ask if he's looking for assistance, go through any necessary training, and get on the volunteer schedule.

The same approach can be used when setting goals for you and your partner. Remember to make the goals attainable and specific, and try not to get too bogged down in creating a step-by-step plan.

When They Launch for Good

No matter how flexible the modern family has become, the time must ultimately arrive for your adult child to leave your nest for good (for the second time) and create a nest of his own. For parents, it can be a bittersweet experience. You're relieved to see your son move on, but you're despondent at the same time. You'll miss having him around—even if he always left dirty dishes in the sink!

Launching your adult child from the nest (even if it is for the second or third time) is one of your life-cycle tasks during middle age. It's a task that involves realigning family roles, like becoming a two-some again with your partner or accepting new family members through marriages or births.

Even though it's developmentally "normal" to launch your adult child, it is an emotional experience. The house may feel lonely or too quiet without your daughter. You may begin to miss your "parent" role. Some moms may cry for a week or two; others are surprised by how easily they adjust. Dads feel the loss, too, especially if they missed some of the child-rearing because of an earlier focus on work.

When you launch your adult children, however, you have so much to look forward to. And all the more so if you didn't give up pursuit of your dreams when your child temporarily rejoined the nest. Now, you'll have more unfettered enjoyment in your life, as you're able to freely pursue your lifelong goals 110 percent! You'll have more happiness in your marriage because you're able to reconnect with the person you fell in love with years ago. The "empty-nest syndrome" is not a syndrome at all. It's a reason to celebrate. Your "open nest" is once again empty—for now, at least.

The Least You Need to Know

◆ Having an adult child return to the nest doesn't mean you have to return to old ideas about who you are; you can still move forward and embrace a new definition of yourself, both as a person and as a parent.

◆ During this life stage, your tasks include things like strengthening your relationship with your partner, developing deeper friendships, and launching your adult child.

◆ Your life is about *being you*. Never forget to value your own unique life journey.

◆ Someday you *will* have an empty nest … we promise!

Appendix A

What Children Will Agree To: A Contract

Not every family will need to form a contract; often a "plan," as we discussed in Chapter 18, will suffice. You should consider using a contract if past experiences with your adult child have shown that he needs extra structure or accountability, or she has trouble taking responsibility or following through.

If your child needs structure, then this contract, in conjunction with your family's plan, will form the foundation for your child's return to the nest. The contract details what specifically your child will agree to, while the plan outlines how your family will achieve its goals.

Reaching Agreement

Though each family's contract will be different depending on its needs, below we offer a boilerplate selection of terms the adult child will want to consider when crafting a contract with her parents. You're encouraged to create a contract that works best for you.

The verbiage is matter of fact, but parents may need to allow for some grace from time to time, such as finding a job by a certain date, or paying off a certain amount of credit card debt. Feel free to pick and choose from the sections we suggest below. You may choose to use only one section and omit the rest. You know your adult child, and you can probably anticipate which areas need to be covered and which are overkill. Keep in mind that the contract exists to keep your adult child focused on the goal of independence.

Financial Obligations

◆ I agree to pay $_____ per month in rent, payable on the _____ day of the month.

◆ I agree to pay $_____ toward utilities every month. *Or* I agree to _____% of the utilities every month.

◆ I agree to purchase my own food.

◆ I agree to pay my own bills, including the following:

— Cell phone

— Car insurance

— Internet access

— Credit card bills

— Tuition

— Other: _____

Around the House

◆ I agree to do the following chores:

— Clean the bathroom once a week

— Take out the garbage every night

— Wash the dishes after every meal I eat at home

— Vacuum the common areas once a week

— Other: _____

- I agree to do my own laundry.

- I agree to keep the common areas neat and tidy, to meet my parents' standards.

- I agree to share the following household electronics:

  — Television

  — DVD player

  — DVR

  — Stereo

  — Computer/home office

  — Video game console

  — Other: _____

- I agree to keep my belongings stored in the garage or _____, and when I move out, I'll take them with me.

Personal Goals

- I agree to find a job by __(this date)__ . *Or* I agree to enroll in school for the _____ semester.

- I agree to move out by __(this date)__ . *Or* I agree to move out as soon as I _____:

  — Find a job

  — Save $_____

  — Graduate from school

  — Pay off my credit card bills/student loan/other debt

  — Find a new apartment

- I agree to be financially independent by __(this date)__ .

- I agree to be clean and sober while living at home.

- I agree to follow house rules regarding smoking.

♦ I agree to attend counseling sessions with a family therapist (or other professional, like a career consultant) once a week or _____ while living at home.

Common Courtesy

♦ I agree to notify my parents if I'm going to be coming home later than planned.

♦ I agree to notify my parents if I'm going to spend the night elsewhere.

♦ I agree to ask permission to use something that does not belong to me, and to abide by agreements regarding its use. (For example, I promise to bring my sister's car back in time for her to drive to work, and to return it with a full tank of gas.)

♦ I agree to notify my parents if I'm going to invite friends over.

♦ I agree to follow house rules regarding having a significant other spend the night.

♦ I agree to participate in __(weekly, monthly)__ family meetings.

What Parents Will Agree To: A Contract

The purpose of the parent contract is not only to solidify financial agreements and responsibilities around the house, but also to acknowledge that your adult child is now an *adult* and will be treated as such. It reinforces changes from the last time your child lived at home.

Not every family will need to form a contract; often a "plan," as we discussed in Chapter 18, will suffice. The adult child may, in fact, be the driver of a contract like this if she feels she shouldn't be the only one held to a contract, or if he has specific concerns about the financial arrangement or level of privacy or autonomy he's entitled to.

Reaching Agreement

Much of this contract will center on how you and your spouse decide to support your adult child. Each family's contract will be different depending on its need, but below, we've listed some terms you may want to consider when crafting your contract with your child.

Financial Allowances

♦ We agree to provide shelter, a room, and all the amenities for ___(this long)___. *Or* We agree to provide space in our house as long as our child follows the terms of her contract.

♦ We agree to give our child $_____ per month as a stipend until he finds a job.

♦ We agree to pay basic household expenses while our child lives at home and other bills as negotiated and agreed at family meetings.

Around the House

♦ We agree to share household appliances and electronics according to the terms of our negotiated family plan.

♦ We agree to let our child store his belongings in the garage or _____ while living at the house.

♦ We agree that the house should be a smoke-free environment.

Common Courtesy

♦ We agree to let our adult child come and go freely.

♦ We agree that all family members will act with respect toward each other.

♦ We agree to give our child complete privacy by not doing the following:

— Entering her room

— Reading or listening to her private communications

— Other: _____

♦ We agree to notify our child if we plan to host a party or event at the house.

♦ We agree to participate in ___(weekly, monthly)___ family meetings.

Appendix C

Resources

Further Reading

Here's a guide to books that we found to be good for deepening your understanding of the family dynamics we introduced in this book.

Parenting and Couples Guides

Burns, David D. *Feeling Good Together: The Secret to Making Troubled Relationships Work.* New York: Broadway, 2008.

Carter, Betty, and Joan Peters. *Love, Honor and Negotiate: Building Partnerships That Last a Lifetime.* New York: Atria, 1997.

Christensen, Andrew, and Neil S. Jacobson. *Reconcilable Differences.* New York: The Guilford Press, 2002.

Coburn, Karen Levin, and Madge Lawrence Teeger. *Letting Go: A Parents' Guide to Understanding the College Years.* New York: HarperCollins Publishers, Inc., 2009

Gottman, John M., and Nan Silver. *The Seven Principles for Making Marriage Work.* New York: Three Rivers Press, 2004.

Kastner, Laura, and Jennifer Wyatt. *The Launching Years: Strategies for Parenting from Senior Year to College Life.* New York: Three Rivers Press, 2002.

Minuchin, Salvador. *Family Healing: Strategies for Hope and Understanding.* Free Press, 1970.

———. *Family Kaleidoscope.* Cambridge, MA: Harvard University Press, 1986.

Olson, David H. *Empowering Couples: Building on Your Strengths.* Minneapolis: Life Innovations Inc., 2000.

Families and Sibling Relationships

Bank, Stephen, and Michael Kahn. *The Sibling Bond (Basic Behavioral Science).* New York: Basic Books, 2008.

Coontz, Stephanie. *The Way We Never Were: American Families and the Nostalgia Trap.* New York: Basic Books, 2000.

———. *The Way We Really Are: Coming to Terms With America's Changing Families.* New York: Basic Books, 1998.

Levitt, Joann, Marjory Levitt, and Joel Levitt. *Sibling Revelry: 8 Steps to Successful Adult Sibling Relationships.* New York: Dell Publishing, 2001.

McGoldrick, Monica. *You Can Go Home Again: Reconnecting with Your Family.* New York: W.W. Norton & Co., 1997.

Overcoming Anxiety and Depression

Bourne, Edmund J., and Garano, Lorna. *Coping with Anxiety: 10 Simple Ways to Relieve Anxiety, Fear & Worry.* Oakland, Calif.: New Harbinger Publications, 2003.

Burns, David D. *The Feeling Good Handbook.* New York: Plume, 1999.

Forsyth, John P., and Georg H. Eifert. *The Mindfulness and Acceptance Workbook for Anxiety: A Guide to Breaking Free from Anxiety, Phobias, and Worry Using Acceptance and Commitment Therapy.* Oakland, Calif.: New Harbinger Publications, 2008.

Goal Setting and Decision Making

Allen, David. *Getting Things Done: The Art of Stress-Free Productivity.* New York: Penguin, 2002.

Davidson, Jeff. *The Complete Idiot's Guide to Getting Things Done.* Indianapolis: Alpha Books, 2005.

Gladwell, Malcolm. *Blink: The Power of Thinking Without Thinking.* New York: Little, Brown and Company, 2005.

Gay and Lesbian Issues

Fairchild, Betty, and Nancy Hayward. *Now That You Know: A Parents' Guide to Understanding Their Gay and Lesbian Children.* New York: Harvest Books, 1998.

Jennings, Kevin, and Pat Shapiro. *Always My Child: A Parent's Guide to Understanding Your Gay, Lesbian, Bisexual, Transgendered or Questioning Son or Daughter.* New York: Fireside, 2002.

McDougall, Bryce. *My Child is Gay: How Parents React When They Hear the News.* Australia: Allen & Unwin, 2007.

Merla, Patrick. *Boys Like Us: Gay Writers Tell Their Coming Out Stories.* New York: Avon Books, 1996.

Retirement and Finance

Carlson, Robert C. *The New Rules of Retirement: Strategies for a Secure Future.* Hoboken, N.J.: Wiley, 2004.

Ernst & Young, Martin Nissenbaum, Barbara J. Raasch, and Charles L. Ratner. *Ernst & Young's Personal Financial Planning Guide.* Hoboken, N.J.: Wiley, 2004.

Fisher, Sarah Young, and Susan Shelly. *The Complete Idiot's Guide to Personal Finance in your 20s and 30s.* Indianapolis: Alpha Books, 2005.

Fletcher, Douglas S. *Life After Work: Redefining Retirement. A step-by-step guide to balancing your life and achieving bliss in the Wisdom Years.* Bangor, ME: Booklocker.com, Inc. 2007.

Tapscott, Don. *Wikinomics: How Mass Collaboration Changes Everything.* New York: Portfolio, 2008.

Wuorio, Jeffrey J. *The Complete Idiot's Guide to Retirement Planning.* Indianapolis: Alpha Books, 2007.

Websites

Though these are just cursory, you can find a wealth of information online. Before you take advice you find on a website, however, make sure you can trust the source fully.

aamft.org   The American Association for Marriage and Family Therapy offers a free therapist locator service for those seeking a licensed family therapist.

aarp.org   The American Association of Retired Persons offers a host of current, relevant information for older generations.

census.gov   The U.S. Census provides statistics information about trends in the United States.

familyeducation.com   Parenting advice for families with infant, adolescent, teenage, and young adult children.

mayoclinic.org   A good resource for health-related research.

ncoa.org   The National Council on Aging lists up-to-date news, events, and lifestyle tips for older people.

Index

A

acceptance
  of child, 131–132
  need for, 31
active listening, 128–129
adolescence, siblings, 140
adulthood, establishing, 200
age differences in siblings,
  138–139
age of marriage, 11
assumptions about other person,
  communication, 130

B

babysitting grandkids, 169–170
bailouts, 159–160
bedroom, 72–73
  as child's space, 73
  compromise, 73
beliefs, 15
benefits of open nesting
  pro-con list
    children, 21–22
    parents, 17–18
birth order differences in siblings,
  139–140
Blink, 14
brainstorming for plan, 192–193
breaks during talks, 134

C

career
  career choice of child, 7
  parents, 28
caring for self, 27
changing child, 131–132
chaotic families, 47, 119–120
childhood memories, young
  adults, 142
children
  acceptance of, 131–132
  career choice, 7
  caring for self, 27
  changing, 131–132
  coming out of the closet, 182
  comparing to one another, 141
  controlling, 163
  differences with parents,
    183–184
  low self-esteem, 96
  never left nest, 11
  not ready to leave, 60–63
  partner moving in, 81–82
    common interests, 84
    communication, 81
    conflict resolution, 81
    enmeshed families, 88
    family meetings, 82
    getting along with family,
      83–84
    intimacy, 86–87
    redecorating, 85
  partners
    multiple, 96–97
    unacceptable, 95–96

pro-con list
   benefits, 21–22
   drawbacks, 23–24
reasons for returning home, 9
regressing, 51
returning home to care for
   parents, 207–208
returning to old roles, 211–212
chores, 18
   child not doing, 106–107
   regressing into old roles, 105
clarity in communication, 48
cleaning kitchen, 75
closeness after experience, 5
cohesion, 44–46, 121
   disengaged families, 123–124
   enmeshed families, 122–123
common areas, 75
common interests with child's
   partner, 84
common space
   guests, 76
   scheduling time, 76
communication
   active listening, 128–129
   asking positively, 130
   assumptions about other
      person, 130
   breaks during talks, 134
   child's partner, 81
   common space usage and, 75
   families moving in, 80
   "I statements," 129–130
   letter writing, 135
   mind reading, 130
   rent, 109
   repair attempts, 131
   sexuality, 91–92, 101
   sleepovers, 95
   timing, 131
   weekly meetings, 135

communication skills, 47, 49
community attitudes, 16, 24
community involvement, 29
comparing siblings to one
   another, 141
conflict-avoidance, space, 70
conflict resolution
   child's partner, 81
   families moving in, 80
   pursuer/chaser, 132, 134
conflicts, 126–127
conformity, 31
connected families, 188
connectedness, 121
connection with child, 17
continuity tracking in communication, 48
contract, moving out, 57
controlling behaviors, 33
controlling persons, 46
cooking, 149–150
costs, added, 21
cultural acceptance, 177–178

D

dating parents, 41, 99–100
   child's disapproval, 101
debt, 10
decision making, 194
democratic leadership, 47
developmental milestones,
   family, 26
differences of opinion, 33
differences with adult children,
   183–184
differentiation from family of
   origin, 27, 31–32
   levels of, 32–33
dining together, 151

discipline
    grandparents, 168–169
    self-discipline, 27
discretionary income, 7
discussion topics, 176
    gay civil rights, 181–183
    politics, 178–179
    religion, 180–181
disengaged families, 123–124
displacement of siblings, 50–51
disputes, 9
division of labor
    child not doing chores,
        106–107
    regressing into old roles, 105
divorce, 11
divorced parent household, 41–42
drawbacks of open nesting
    pro-con list
        children, 23–24
        parents, 19–21
dropping everything for child,
    210–211
dry-erase board, 149
dynamics of the family, 49–51

E

eating together, 151
electronics
    scheduling time, 78
    sharing, 76–77
emotional adjustments, 116–117
emotional support for child,
    206–207
emotional support of child, 22
enmeshed families, 88, 122–123
entitlement, 161
executive system
    among parents, 50
    couple as, 163
exercises, goal-setting, 37–38

F

families
    connected, 188
    flexible, 188
    separated, 188
    structured, 188
families moving in
    child's partner, 81-82
    communication, 80
    conflict resolution, 80
    enmeshed families, 88
    family meetings, 82
    getting along with partner,
        83–84
    intimacy, 86–87
    redecorating, 85
    similarities to partner, 83–84
family
    chaotic, 47
    cohesion of, 44-46
    communication skills, 47-49
    developmental milestones, 26
    flexibility, 46–47
    functions, 44
    rigid, 46
    strength as family unit, 18
    structured, 47
    traditional family, 40–41
    types, 40
        multigenerational, 43–44
        remarried parent, 42–43
        single/divorced parent,
            41–42
family dynamics, 49-51
family goals, 6
family history, 15
family meetings, 135, 189, 195
    child's partner, 82
    guidelines, 194, 196–197

family of origin
attitudes, 15
differentiation from, 27
family style, planning,
188–189
finances, 18
bailouts, 159–160
controlling child with, 163
enabling child, 160–162
expectations, 157
family history, 158
food bill, 148
retirement, 156
teaching child about, 158–159
tension in couple, 162–163
financial support of child, 21
flexibility, 118
chaotic families, 119–120
rigid families, 118–119
flexibility of family, 46–47, 188
flooding during talks, 134
food bill, 148
grocery shopping, 149
meal planning, 148–150
food preparation, 149–150
dining together, 151
friendships, parents, 29
functions of a family, 44

G

gay child coming out of the closet,
182
gay civil rights discussions,
181–183
Gladwell, Malcolm, 14
goals
achieving, 34–36
conflicting spousal goals on
moving out, 58, 60

life goals, 26
parent-child, mismatched,
56–58
goal setting, 27, 34–36
example situation, 36–37
exercise, 37–38
staying on track, 214–215
goals of family, 6
grandchildren
triangulation, 166
working with grandkids, 174
grandparents, 166
babysitting, 169–170
creativity with grandkids, 173
discipline, 168–169
excursions with grandkids, 173
feedback, 168–169
life-cycle tasks, 167
physical precautions, 174
raising grandkids, 171
adjustment period, 172
testing period, 172
relationship with grandchild,
166
sharing history with grandkids,
173
video games with grandkids,
174
Gray, Lauren, 4–5
grocery shopping, 149
guests, warning, 76
guilt feelings by parent, 213

H

half-siblings, 146
head of household, 120
health connectedness, 121
housing costs, 7, 10

I

"I statements" in communications, 129–130
income, discretionary, 7
independence, 26
asserting, 30
establishing, 30–31
individuation, 31–32
influence of parents, 23
instincts about situation, following, 14
intimacy, 86–87. *See also* sexuality

J–K

job as parent, 17
job creation, 10
jobs, skill requirements, 10

kitchen, 73–75
clean-up, 75
missing food from fridge, 74
separate shelves, 74

L

ladder climbing, 11
launched siblings, one at home, 144–145
launched siblings and returning siblings, 143–144
launching for good, 215
Leave It to Beaver, 40
length of stay, 56, 58
child not ready to leave, 60–61, 63
conflicting spousal goals, 58–60
parental pity, 66–67
parental road blocks, 64–65
reasons for moving, 57–58

letter writing in conflict, 135
life-cycle tasks, 26
continue working toward, 50
grandparenthood and, 167
parents, 28–29
young adults, 27–28
life goals, 26
life on hold, 19
listening, active, 128–129
listening in communication, 48
living space, 75–76
scheduling time, 76
lost path, child, 204–205
low self-esteem in child, 96
lower incomes, 10

M

maintaining routine, 50
marriage, triangulation, 20
marriage age, 11
meal planning, 148–150
mental adjustments, 116–117
mental health issues, 61-63
middle child characteristics, 140
mind reading, 130
missed rent payments, 110
missing food from fridge, 74
money management, 156
controlling child with money, 163
enabling child, 160–162
expectations, 157
family history, 158
retirement money, 156
teaching, 158–159
tension in couple, 162–163
moving out
child not ready to leave, 60–61, 63
contract, 57
parental pity, 66–67

parental roadblocks, 64–65
planning for, 193–194
reasons for, 57–58
multi-media hub, 77
multigenerational households,
43–44

N–O

negative judgments, 15
never left nest, 11
nurturing
developing, 28
parents, 28

old roles, slipping back into, 51–52
oldest child, characteristics, 139
Ozzie and Harriet, 40

P

parent-child bond, 22
parent-child goals, mismatched,
56–58
parental pity for child, 66–67
parental roadblocks to child leaving, 64–65
parents
career, 28
community involvement, 29
conflicting goals on length of
stay, 58-60
dating, 41
child's disapproval, 101
dropping everything for child,
210–211
emotional support for child,
206–207
executive system, 50, 163
friendships, 29
guilt feelings, 213

head of household, 120
influence on children, 23
life-cycle tasks, 28–29
needed child to return,
207–208
nurturing, 28
paying for housing, 7
pro-con list
benefits, 17–18
drawbacks, 19–21
psychological need to keep
children home, 7
returning to old roles, 211–212
sexuality
child's disapproval, 100–101
dating, 99–100
third-wheel feeling, 49–50
triangulation, 65
parents' willingness, 7
partner of child, 81–82
common interests, 84
communication, 81
conflict resolution, 81
family meetings, 82
getting along with family,
83–84
moving in
enmeshed families, 88
intimacy, 86–87
redecorating, 85
multiple, 96–97
unacceptable, 95–96
pity for child, 66–67
planning, 187
brainstorming, 192–193
family style, 188–189
officialness of plan, 189
questions to ask, 8, 190–191
re-launch, 193–194
rigid style families, 189
planning for progress, 58
planting roots, child, 205–206

points of view, discussion topics, 176
  gay civil rights, 181–183
  politics, 178–179
  religion, 180–181
political discussions, 178–179
positive attitude, 135
privacy, 7
privacy issues, 24
pro-con list, 15
  children
    benefits, 21–22
    drawbacks, 23–24
  creating, 16–17
  parents
    benefits, 17–18
    drawbacks, 19–21
progress plan, 58
psychological needs of parents, 7
pursuer/chaser in conflict, 132-134
putting life on hold, 19

Q

questions to ask during planning, 8
questions to ask for plan, 190–191

R

racial tolerance, 177–178
re-launch, planning for, 193–194
reasons for returning home, 9
red flags, 11–12
redecorating child's room, 85
regressing children, 51
regressing into old roles, 105
relationships
  life-cycle tasks, 28
  sociologists on, 7

religious discussions, 180–181
remarried parent household, 42–43
rent
  amount to charge, 109
  charging later, 111
  communication, 109
  deciding to charge, 108
  missed payments, 110
repair attempts during disputes, 131
repeated returning, 203
  paths lost, 204–205
  planting roots, 205–206
  reasonableness, 201
  transitions, 202–204
requesting positively, 130
resentment, 20
respect, communication, 49
response to child's presence, 9
restraint of children, 23
retirement money, 156
returning to old roles, 211–212
rigid families, 46, 118–119
  plan, 189
routine, maintaining, 50

S

scheduling electronics time, 78
scheduling time for use of common space, 76
self-discipline, 27
self-disclosure in communication, 48
separated families, 188
sexuality, 27. *See also* intimacy
  attitudes toward, 91
  communication, 91–92, 101

parents
    child's disapproval, 100–101
    dating, 99–100
    sleepovers, 94–95
        younger siblings, 98
    sleepovers away, 92–94
sharing resources with siblings,
    142
sharing space, 20
siblings
    adolescents, 140
    age differences, 138–139
    birth order differences,
        139–140
    childhood memories, 142
    comparing to one another, 141
    competitiveness, 142
    displacement of, 50–51
    half-siblings, 146
    launched and returning,
        143–144
    one home, one launched,
        144–145
    sexuality, 98
    sharing resources, 142
    step-siblings, 146
    support for each other, 142
    young adults, 141–142
single/divorced parent household,
    41–42
sleepovers, 94–95
    younger siblings, 98
sleepovers away, 92–94
space
    bedroom, 72–73
        compromise, 73
    changes in needs, 71
    common areas, 75
    conflict-avoidance, 70
    electronics, 76–77
    guests, 76
    kitchen, 73–75

living space, 75–76
need for, 70–71
technology, 76–77
space sharing, 20
speaking in communication, 48
standard of living, 10
staying on track, goal setting,
    214–215
step-siblings, 146
strained relationships, 23
strength as family unit, 18
structured families, 47, 188
subjects of discussion, 176
    gay civil rights, 181–183
    politics, 178–179
    religion, 180–181

T–U–V

technology
    scheduling time, 78
    sharing, 76–77
tension over finances, 162–163
tensions, 126–127
third-wheel feeling of parents,
    49–50
time frame for stay, 193
time period of stay, 56-58
    child not ready to leave, 60–61,
        63
    conflicting spousal goals, 58,
        60
    parental pity, 66–67
    parental roadblocks, 64–65
    reasons to move out, 57–58
timing of discussions, 131
traditional family, 40–41
traditions, 15
transitions, repeated returning,
    202–204
triangulation, 20, 65
    child's partner and, 81
    grandchildren, 166

W

willingness of parents, 7
worrying about child, 19

X-Y-Z

young adults,
    life-cycle tasks, 27–28
    siblings, 141–142
younger siblings, displacement of,
    50–51
youngest child, characteristics,
    140

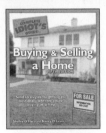

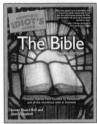

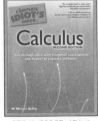